SPECIES CONCEPTS AND PHYLOGENETIC THEORY

SPECIES CONCEPTS AND PHYLOGENETIC THEORY

A DEBATE

Edited by Quentin D. Wheeler and Rudolf Meier

COLUMBIA UNIVERSITY PRESS NEW YORK

Columbia University Press
Publishers Since 1893
New York Chichester, West Sussex

Library of Congress Cataloging-in-Publication Data
Species concepts and phylogenetic theory : a debate / edited by Quentin D. Wheeler and Rudolf Meier.
p. cm.
Includes bibliographical references.
ISBN 0-231-10142-2 (cloth : alk. paper) — ISBN 0-231-10143-0 (paper : alk. paper)
1. Species. I. Wheeler, Quentin, 1954–. II. Meier, Rudolf.
QH83 .S64 2000
576.8'6 21—dc21 99-044163

Casebound editions of Columbia University Press books are printed on permanent and durable acid-free paper.
Printed in the United States of America
c 10 9 8 7 6 5 4 3 2 1
p 10 9 8 7 6 5 4 3

CONTENTS

PART 2. CRITIQUE PAPERS (COUNTERPOINT)

PART 3. REPLY PAPERS (REBUTTAL)

CONTRIBUTORS

Dr. Joel Cracraft
Department of Ornithology
American Museum of Natural History
Central Park West at 79th Street
New York, New York 10025, USA

Dr. Richard L. Mayden
Department of Biological Sciences
Biodiversity and Systematics
University of Alabama
Tuscaloosa, Alabama 35487, USA

Dr. Ernst Mayr
Museum of Comparative Zoology
Harvard University
Cambridge, Massachusetts 02138, USA

Dr. Rudolf Meier
Zoological Museum
Universitetsparken 15
2100 Copenhagen 0, Denmark

Dr. Brent D. Mishler
Department of Integrative Biology, Jepson Herbarium, and University Herbaria
University of California at Berkeley
Valley Life Sciences Building
Berkeley, California 94720, USA

Dr. Norman I. Platnick
Division of Invertebrate Zoology
American Museum of Natural History
Central Park West at 79th Street
New York, New York 10024, USA

Dr. Edward C. Theriot
Section of Integrative Biology
and Texas Memorial Museum
University of Texas
Austin, Texas 78712, USA

Dr. Quentin D. Wheeler
Department of Entomology and L. H. Bailey Hortorium
Cornell University
Comstock Hall
Ithaca, New York 14853, USA

Dr. E. O. Wiley
Natural History Museum and Department of Ecology and Evolutionary Biology
The University of Kansas
Lawrence, Kansas 66045, USA

Dr. Rainer Willmann
II. Zoologisches Institut
Universität Göttingen
Berliner Straße 28
37073 Göttingen, Germany

PREFACE

The rapid rise of phylogenetic theory since Hennig's seminal 1966 book has at an unprecedented pace changed the way that systematists and taxonomists do their work, as well as the quality of their hypotheses and classifications and their utility to all biology. Because species occupy a pivotal position in all aspects of biology in general and phylogenetic systematics in particular, it is critically important that the concept of species be compatible with these profound advances in phylogenetic theory. To this scientific significance, add also a growing awareness of the potential for mass species extinctions in the immediate decades ahead (e.g., Wilson, 1985, 1992) and the dire need for changes in conservation biology that minimize negative impacts of the "biodiversity crisis" while conserving as much biological diversity as possible. Even the simplest scientific responses to the biodiversity crisis, such as establishing what kinds and how many organisms live on planet earth or comparing the relative diversity of two taxa or areas, depend in no small measure upon general agreement about what a species is. Surprisingly, and in spite of literally thousands of scientific papers relevant to the subject, there are more species concepts in popular usage today than at any point in the past century, and the consensus in zoology about the Biological Species Concept has begun to unravel. An aggressive search for a species concept that is consistent with phylogenetic theory has begun.

This volume evolved from long-term and unresolved differences of opinion with regard to the nature of species between the two co-editors. After many fruitless and sometimes loud discussions, we thought it desirable to expand such argumentation to include several additional concepts of species prevalent in contemporary biological literature. It was our belief that a face-to-face debate among proponents of the various concepts was likely to produce more heat than light, and that a "virtual" debate format that would combine the point/counterpoint advantages of a debate with the dispassionate composition of statements in the comfort of one's own office and in the presence of the literature resources that can back up positions was preferable.

We hoped that such a book would also fill a void in the species concept literature. Normally, different species concepts are introduced back to back, and it remains unclear what the main differences between the species are, why the authors reject each others' concepts, and, most of all, how they would respond to criticism of the more or less obvious shortcomings of their own concept. Our debate format has forced authors to confront these questions.

Columbia University Press was enthusiastically receptive to our idea for such a written debate, and this book is the result. The path to this end has been long and tortuous, much like the recent history of species concepts. The project, initiated in 1993, was logistically difficult, requiring three subsequent essays from all authors. This lengthy process explains the absence of the most recent literature. However, the most important points of departure between the concepts and the most critical issues have changed little, and the need for this debate has not lessened. In order to read and follow the debate that follows, it may be useful for the reader to understand a bit more about the debate's format and rules of engagement.

In order to make the debate format possible, the book was organized into three sections. In the first set of essays, the authors were given the platform to prepare position papers, defining their respective concepts and making opening remarks designed to convince the reader of the merits of their particular point of view. The second set forms critiques of the four competing concepts—the counterpoint of the debate. The third and final set are responses or rebuttals to the attacks launched in the second set of essays. The best assurance of our objectivity as editors is the fact that we still disagree with each other to the same or greater degree than either of us does with the other contributors. Thus, we have kept each other honest.

We have already received some criticism on our selection of or limits on those invited to participate in this debate. There are, to be sure, additional species concepts that might have been included. However, we determined to focus the debate on those concepts that have received the most serious consideration by phylogenetic systematists. There are many books on species concepts: some present various concepts; others discuss how the choice of concept might affect the study of speciation; still others focus on taxon-specific differences. Strangely enough, a book was missing that discussed the choice of species concept for phylogenetic systematists and cladists, an ever-growing and influential group of scientists. In the end, we determined to invite the following proponents, a diverse group of authors representing botany, zoology, and paleontology, to present their respective species concepts: Ernst Mayr (the Biological Species Concept); E. O. Wiley and Richard Mayden (the Evolutionary Species Concept); Rudolf Meier and Rainer Willmann (the Hennigian Species Concept); Brent Mishler and Edward Theriot (their version of the Phylogenetic Species Concept); and Quentin Wheeler and Norman Platnick (with a competing version of the Phylogenetic Species Concept).

An introductory essay was solicited from Dr. Joel Cracraft, who, as one of the earliest outspoken proponents of a phylogenetic species concept, has a very long and active involvement in the community's debate over species concepts, and is in an ideal position to explain the historical and contemporary philosophical and theoretical contexts that make such a debate necessary and desirable. Dr. Cracraft's essay was written after the debate was over so that he would have the benefit of having read the arguments in their entirety before setting the stage for the printed debate.

We apologize for any potential confusion in the text regarding the use of the phrase *Phylogenetic Species Concept*, but two major competing views have emerged, each vying for the ultimate claim to that title and its explicit verbal relationship to phylogenetics. In an attempt at fairness, we have permitted the authors to use such terms as they see fit. Where we, as editors, have perceived the possibility of any confusion, we have used the modifiers *sensu Mishler and Theriot* for the Mishler and Theriot version and *sensu Wheeler and Platnick* for the Wheeler and Platnick version of the phylogenetic concept. In general, however, the risk of such confusion is greatly diminished by simply keeping in mind who wrote the particular passage in question.

Although Cracraft's essay explains why such a debate is needed more than a century after the general acceptance of Darwin's explanation for the origin of species, nowhere in the book is a winner in the debate identified. This will be left initially to the reader and, ultimately, to the adoption of these competing concepts by fellow biologists. The stakes are high, and the use of the "wrong" species concept has the potential to create enormous and long-term problems for basic biological research, for assessment of biological diversity, and for progress in conservation efforts. The question of species is fundamental to biology as a whole, and we must endeavor to arrive at a consensus—the right consensus—as quickly, efficiently, and accurately as possible. We hope that this debate will be the first step toward that goal.

As the editors, we owe a great debt to a considerable number of people who have made this volume possible. First and foremost, we thank Dr. Cracraft for providing a conceptual and historical overview of the status of species concepts and the need for a debate, and the debators who have delivered manuscripts with good humor and a degree of mutual respect, only occasionally strained by cutting prose. We thank Ed Lugenbeel and the staff of Columbia University Press for believing in the value of our idea for a written debate and for saintly patience as we worked through what became a very long and laborious process. Finally, we thank all those colleagues who have entered into discussions, verbal debates, and heated arguments with us (individually and together) about species. At various times and venues, such discourse has included graduate students and faculty in the Department of Entomology and the L. H. Bailey Hortorium at Cornell University, colleagues at professional meetings and during visits to their institutions, and students enrolled in courses where such matters were discussed.

Although we still disagree about the best concept of species, we agree heartily about the serious and pressing need to address and resolve the "species problem." As scientists and citizens of a world entering an age of environmental crises and biodiversity challenges, we believe that the question of species concepts is more relevant and important than ever and is of far greater significance than a clashing of intellectual positions. We hope that this volume will find a place in that community discourse, as a book read by student and professional biologists and, more important, read in formal and informal discussion groups where these ideas can be debated in person by the participants.

SPECIES CONCEPTS AND PHYLOGENETIC THEORY

Introduction

1

Species Concepts in Theoretical and Applied Biology: A Systematic Debate with Consequences

Joel Cracraft

Biologists, especially systematists, had debated species concepts for a very long time, well into the nineteenth century. The debate intensified with the rise of the so-called New Synthesis in the 1930s and 1940s and then accelerated even more with the "systematics wars" of the 1960s to 1980s, particularly with the ascendancy of cladistics, or phylogenetic systematics.

The debate itself has had many nuances. Some systematists have only been interested in discriminating all the discrete taxonomic variation they can in nature, without concern for the processes that might have produced this variation. Often called alpha taxonomists, they have been the workhorses of taxonomy, relentlessly monographing Earth's diversity. Although their propensity to describe species has been belittled by some, without their efforts we would know far less about the diversity of the natural world. In the early decades of the twentieth century, a number of systematists, mostly European, became more interested in how this taxonomic variation might have been generated, and they perceived the application of species status to discrete populations as an impediment to these efforts, especially when that variation was deemed minor. These systematists, mostly vertebrate zoologists working on birds and mammals, promoted a "polytypic species concept," which evolved into the well-known Biological

Editors' Note: This introductory essay was solicited from Dr. Cracraft, an innovator and major proponent of a phylogenetic concept of species. It explains why a debate about species concepts is needed today, after so many years and publications devoted to the subject. However, anyone with such in-depth expertise on species concepts necessarily brings opinions and biases to any essay on species, whether explicit or implied. In fairness to all contributors to this volume, the reader is thus cautioned to keep in mind that certain statements in the introduction might be controversial. Therefore, all the essays in this volume, including the introduction, should be read with a critical mind.

Species Concept. Within the framework of the polytypic concept, evolutionary history—speciation in this context—was primarily deduced by a process of lumping phenotypically similar allopatric populations into one geographically widespread species.

These two approaches to species—the traditional alpha-level description of species and the application of a polytypic species concept—pretty much existed side by side throughout most of the succeeding decades, and, at least retrospectively, the debate seemed to quiet down. It picked up again in the 1960s with the introduction of numerical phenetics (numerical taxonomy). Although proponents of the latter philosophy did not have a specific axe to grind about how species should be conceived (Sneath and Sokal 1973), they were interested in building branching diagrams and needed terminal taxa in their analysis. To them, species were another "operational taxonomic unit," albeit one that tended to be the smallest distinct cluster of organisms.

Numerical taxonomy waned for many reasons, but primarily because more and more systematists adopted phylogenetic systematics, or cladistics. Cladistics, like numerical taxonomy, also focused on branching diagrams; but compared with numerical taxonomy, cladistics embraced a more formal framework for examining the history of life, and eventually that extended to a more formal consideration of just what it meant to be a species and what the role of species was in phylogenetic analysis. It becomes clear upon reading the cladistics literature, beginning with Hennig, who leaned toward something akin to a biological species concept, to the writings cited by the authors in this book and their own contributions here, that there is no generally accepted species concept within cladistics. Although most of the cladists represented in this book generally agree on the methods used for reconstructing phylogenetic history, their assertions in their papers suggest that they do not agree on what the lowest-level terminals of those cladograms should be.

The debate in these pages reflects, possibly, why it has been so difficult to find common ground over species concepts throughout the history of systematics. Several reasons for this situation can be suggested (see also Hull 1997):

1. First, the refrain: "Variation in my organisms is not partitioned in nature like variation in your group; hence, how I apply a species concept must be different." Life's diversity is indeed incredibly variable and complex. Specialists who work on, say, liverworts apparently do not see variation in their organisms partitioned into the same types of patterns that someone who works on, say, birds. Moreover, it is often claimed that asexual organisms are so different from bisexual organisms that they cannot be included within the same kind of species concept. And so on.

There has been something of a historical relationship between an adopted species concept and the taxonomic group being studied (papers in Claridge et al. 1997). Thus, for many decades now, ornithologists, mammalogists, and specialists from a few other disciplines have generally adopted a Biological Species Concept; most invertebrate zoologists, on the other hand, including the vast majority of systematists, have largely been indifferent to the Biological Species Concept in their day-to-day work and instead have tended to apply species status to patterns of discrete variation. Botanists have been somewhere in the middle, although most have not used a Biological Species Concept.

So systematists, rightly or wrongly, have believed that the characteristics of their organisms—life histories, patterns of variation, extent of diversity—have something to do with the kind of species concept that must be adopted. And this thinking, this apparent complexity and confusion, also has led some systematists to call for *pluralism*, that is, for the general acceptance that systematics must be willing to embrace many different species concepts.

2. Systematists also have differed in their interest in exploring what might be termed the "meaning" of species. Some have eschewed any concern for meaning and looked upon species as vehicles to describe variation and taxonomic diversity. Others have sought meaning: species as the units of evolution, as the products of evolution, as incipient species, and so on. These different approaches to species have influenced the choice of species concepts, although not in any straightforward way. But, as will be discussed below, there has been a keen desire on the part of many systematists, even those not sharing the same view of species, to find a concept that fits those "units of nature" that we theoretically expect processes of evolution (diversification) to have produced but that is also consistent with our methods for retrieving evolutionary-phylogenetic history. Because biologists see all these parameters in different ways, they frequently see species in different ways.

Mayden (1997) has identified at least 22 species concepts within the contemporary literature. One might take issue with his classification and description, but he has demonstrated that systematists have found ways to conceptualize species to a point that numbs the mind. Not all of these species concepts are significantly different from one another, of course, but many are, and as we shall see, many consequences follow.

The question thus arises: Do we expect the species problem to be resolved anytime soon? If one is to judge by the essays in this book, the answer would seem to be no. With one exception (Mayr), all the contributors to this book profess to be cladists or phylogeneticists, yet they seemingly cannot agree on how species should be perceived. The value of the essays is that they highlight many of the reasons why it has not been possible to settle on a common concept of species. Because many of the arguments involved in this debate are subtle, this introductory chapter is a guide of sorts. All discussions about species should be approached with skepticism, with a critical mind for the nuances of language, of debating ploys, and an appreciation that arguments and conclusions, while using the same words, might not mean the same thing because those words imply different things to different people and because people argue from different premises. Keeping all of this in mind will be the only way one can begin to make sense of these debates. Even then, one will probably need to grab a favorite fetish and conjure up a bit of luck. Lacking a fetish, this introductory chapter may just be the ticket.

Species Concepts Matter

What good are definitions anyway? Why should a debate about species be taken seriously? These are questions many biologists ask out loud when they become frustrated with all the hyperbole surrounding species concepts. Many biologists, but mostly nonsystematists, are quite comfortable with the concept they learned when they took introductory evolutionary biology in college and see no need to change. After all, their species concept seemingly fits what they do—population biology, behavior, genetics, or ecology. The trouble is, they often do not realize that virtually any species concept will be simpatico with what they do because what they do is generally not comparative and involves populations and/or species at a specific point in space and time. But if one is interested in comparing taxonomic entities, or wants to seek a historical interpretation of data, then species concepts matter.

Plenty has been written about why different species concepts "impose" different interpretations on the biological world. The primary reason for being concerned about species definitions is that they frequently lead us to divide nature in very different ways. If we accept the assumption of most systematists and evolutionists that species are real things in nature, and if the sets of species specified by different concepts do not overlap, then it is reasonable to conclude that real entities of the world are being confused. It becomes a fundamental scientific issue when one cannot even count the basic units of biological diversity. After one reads the essays in this book, it should be readily apparent that the widespread use of different species concepts is a confounding influence on describing biological diversity.

Individuating nature "correctly" is central to comparative biology and to teasing apart pattern and process, cause and effect (Cracraft 1989a). Thus, time-honored questions in evolutionary biology—from describing patterns of geographic variation and modes of speciation, to mapping character state or ecological change through time, to biogeographic analysis and the genetics of speciation, or to virtually any comparison one might make—will depend for their answer on how a biologist looks at species.

The importance of species concepts is not restricted to the seemingly arcane world of systematics and evolutionary biology. They are central to solving real-world practical problems that affect people's lives and well-being. As one reads these essays, it is instructive to push aside all the theoretical rhetoric and argumentation and ask how alternative concepts might play out in the real world. Consider, for example, cases in which species concepts might have important consequences: (1) a group of nematodes that attack food crops, or act as vectors for plant viruses, where failure to individuate species correctly might mean that food supplies are at risk (Hunt 1997); (2) a group of exotic beetles that attack timber resources, where failure to individuate species correctly might mean that their place of origin could be misidentified and thus potential biological control agents overlooked; or (3) a group of rodents or insects, where failure to individuate species correctly might mean that a disease vector could be misidentified, thus jeopardizing control programs (Lane 1997); and on and on and on.

Systematics is the fundamental science of biodiversity (Systematics Agenda 2000 1994a, 1994b), and species are arguably systematics' elementary particles. There are practical consequences to every species concept if those elementary particles are not discovered and understood properly. More and more systematists recognize this, as attested by the remarkable increase in concern about how different species concepts are affecting the business of conservation (Ryder 1986; Cracraft 1991, 1997; Geist 1992; Moritz 1994a, 1994b, 1995; Vogler and DeSalle 1994; Grant 1995; Barrowclough and Flesness 1996; among many others). Reading the essays in this volume and then thinking about how species concepts affect (1) the specific status of diagnosable populations, (2) estimates of species diversity, (3) the historical analysis of these units, (4) an understanding of patterns of gene flow within and among these units, (5) delineation of areas of endemism, (6) the demographic characterization of such units, (7) decisions on captive breeding (how much space is devoted to each unit; which unit can be bred with which other unit), and (8) which units will be given protection under local, national, or international legal instruments should be sufficient to drive the point home that species concepts matter (Barrowclough and Flesness 1996; Cracraft 1997).

Similarities: All Else Is Rhetoric

The literature on species concepts is riddled with confusion and obfuscation; indeed, much of the contention seemingly derives from the efforts of advocates of a particular concept to paint their opponents into a corner by creating a caricature of their concept, even when that caricaturization is empty of any significant content. A student of species concepts must be able to sort through the rhetoric, unless, of course, the goal is to use it for one's own gain. In fact, virtually all species concepts—certainly those discussed in this book—are remarkably similar in many ways, and one would think those similarities would be obvious, to the point where the rhetoric would not flow so readily. So let us add some viscosity to the debate.

TYPOLOGICAL/NONPOPULATIONAL

To be typological means that one sees variation compartmentalized into idealized types, that individual specimens, in this case, possess some essential character or form (an *essence*) conferring species status, and that populational variation does not matter. Judging by the number of times this descriptive moniker appears in the literature, the field of systematics must be rampant with systematists disregarding populational variation and settling on one or a few specimens as the basis for individuating species units in nature.

The literature of systematics, however, would suggest that this is not true. Virtually all systematists over the past century have looked on species from a populational viewpoint. For example, would a systematist knowingly describe males and females as separate species, or juveniles as a species separate from adults, merely because they are different? To be sure, many systematists have made mistakes, or have had inadequate information on reproductive behavior and relationships, and often only one sex has been available at the time of a species' description. Systematists work with the materials and knowledge at hand, and some are more careful or perceptive than others. But this does not mean they are typological and willfully ignore variation. So it is important when seeing the word *typological* to inquire about its context and to examine how a systematist might be interpreting populational variation when drawing species limits.

MORPHOLOGICAL

Parallel to claiming that a concept is typological is to say it is "merely morphological," or something to that effect. Again, this argument needs careful examination. Even though a species definition may or may not make reference to characters—which may or may not mean morphological characters—all concepts use character data to adjudicate species boundaries, and most of those data are morphological. Even in the most extreme situations, such as in entomology or paleontology, where the discovery of new species is often based on single specimens, the systematist is not ignoring variation when applying a species definition. If the specimens were there, variation would be taken into account. This has nothing to do with typological thinking.

BIOLOGICAL

One sometimes sees the claim that a concept is nonbiological. Yet it would seem that all species concepts are biological—they are trying to bring some order to life. Moreover, one rarely, if ever, has an advocate of this argument telling the reader exactly what is meant by *nonbiological.*

EVOLUTIONARY AND GENETIC

The same is true for the claim that a concept is nonevolutionary or nongenetic. Simply because a definition does not contain words such as *evolutionary, genetic, lineage, reproduction,* or the like does not mean these are not part of the conceptual underpinning of the concept. All species concepts are genetic (whatever that may mean in this context), and all species are somehow conceived as lineages that have evolved. How could it be otherwise?

SPECIES CATEGORY VERSUS SPECIES TAXON

This dichotomy of language may or may not be understandable, depending on how much basic systematics one has learned. Certainly, it is not a problem for systematists. Most evolutionary biologists probably understand the distinction between the species category within the Linnaean hierarchy of taxonomic categories and the taxon (group) that is ranked at the species level within that hierarchy. Any confusion in the literature is usually the result of a lapse in language and rarely leads to consequences that are significant.

LINEAGE

Some species concepts are characterized as being lineage concepts. Yet, all systematists see the history of life as diversifying lineages, and all species concepts somehow are trying to capture that diversification, whether or not they use this type of language in their definition. To this extent, all species concepts are similar.

ARBITRARY, SUBJECTIVE, AND OBJECTIVE

Calling a species concept *arbitrary* or *subjective* seems primarily to be a debating ploy even though one might really believe that some species boundaries are being individuated in a nearly arbitrary manner. The implication is that one's own concept has pride of place in being objective.

Virtually all systematists taking part in the species debate, including myself, have used the word *arbitrary* or *subjective* when discussing one or more species concepts. The truth of the matter is that systematists are rarely, if ever, indisputably arbitrary; it is not as if they look at patterns of variation in nature and then use a random-numbers table to draw boundaries of taxa. They bring to bear on the interpretation of that variation their prejudices, experiences, the data available, and the theoretical framework of the species concept within which they operate—hardly arbitrary or subjective. Nor is any concept entirely objective for somewhat the same reasons. In the real world, available data are often ambiguous—there are never enough specimens, or they are not appropriately distributed in space or time—so even if one knows what *objective* is supposed to mean in these situations, one frequently makes inferences that are not clear-cut. All systematists do this, no matter what their species concept.

My Concept Is the Best

A consistent refrain in the literature on species concept is that own's own concept is best for addressing the taxonomic diversity of nature. Has that changed with the authors of this book? Consider:

The Biological Species Concept is best: "To be sure, assigning populations to biological species on the basis of the criteria discussed by Mayr (1969:181–187) will not eliminate the possibility of an occasional mistake. However, it [the Biological Species Concept] is the best method available to a biologist" (chapter 2).

The Hennigian Species Concept is best: "In our opinion, the Hennigian Species Concept is currently the best species concept for bisexual organisms" (chapter 13).

The Phylogenetic Species Concept (*sensu* Mishler and Theriot) is best: "Our Phylogenetic Species Concept, with its explicit acknowledgment of the need for application of different ranking criteria in different cases, lends itself to a more rational assessment of biodiversity, one lineage at a time instead of a mindless counting of species names" (chapter 14).

The Phylogenetic Species Concept (*sensu* Wheeler and Platnick) is best: "[O]ur version of the Phylogenetic Species Concept is more useful in both a taxonomic and general biological sense and should be adopted over the other phylogenetic concept as well as others in this book" (chapter 15).

The Evolutionary Species Concept is best: "[T]he Evolutionary Species Concept is the only concept currently capable of recognizing all naturally occurring biological taxonomic entities" (chapter 6).

No rapprochement here. Everyone seems to believe that his concept is best. But the notion of "best" is always relative, of course. The reader is encouraged to look hard at the context of what *best* might mean. Does it mean only bisexual organisms or does it apply to all kinds, sexual and asexual? Thus, how restrictive taxonomically is a particular concept? Does *best* mean the concept can be applied throughout space and time? Again, is the concept restricted in its applicability to the fossil record as well as to extant organisms, or to allopatric, parapatric, or sympatric situations? What we are exploring here is whether the concept is general, and the reader will have to ask whether a general concept is more useful ("best") than one that is less general. A lot of question marks, yet the questions are not without relevance.

How Do We Know What We Think We Know?

The literature on species concepts has always contained a high proportion of philosophical ruminations by biologists and has received considerable attention by philosophers. The biologists generally are not very sophisticated as philosophers and the philosophers are not very sophisticated as biologists—but that is all right. At least there has been a dialogue, and it has sharpened and informed the discussion and controversy in useful ways.

Take the notions of reality, individuality, and entities. All of these figure prominently in recent discussions about species concepts, and the essays in this book are no different. All the authors in this volume are realists—you will not find them talking about fictitious things. But one has to exercise caution because as with all discussions about reality, it always pays to be aware of my sixth law of thermodynamics, never falsified: things aren't as they seem.

When systematists talk about species, they almost always think of them as being real entities. It is worthwhile asking what this means to the person making this claim. To be a real entity, something should have discrete boundaries in space and time, but it must be remembered that boundaries are fuzzy or discrete, depending on the spatiotemporal point of view of the "observer," using either one's own senses or some device designed for observing (Cracraft 1987, 1989b). We could put 20 biologists in a room and point to a squirrel, and everyone would probably agree that, yes, that is a real entity, a real individual—a squirrel. But one cannot do that with a species. Biologists do not see species, yet that does not prevent us from inferring they are discrete, real entities. One of the things that helps us with that inference is a definition, which forms the basis for individuating what we think are real entities called species.

There is no doubt that if biologists concoct a definition of species, those things individuated by that concept are real to them. But there is a kicker in all this: definitions do not necessarily make things real, and some definitions might actually lead us to identify spurious ("unreal") entities. In what sense might this be true? By saying that something is a real discrete entity, we imply that it participates in one or more natural processes. Individual organisms participate in a host of processes: they predate, escape from predation, eat, and so forth. On the other hand, some processes are best considered as involving entities other than the entire organism itself (e.g., absorbing nutrients, which is done by cells). Species, likewise, are thought to participate in processes. Some of these, however, do not involve species as discrete entities. For example, species do not compete for resources because individual organisms can be said to do that; species also are not participants in processes such as extinction in the sense of "going extinct" rather than "becoming extinct." Nitpicking, one might say, but if that is the case, then one could argue that there is no need to talk about birth processes of individual organisms not keeping up with death processes when speaking of causes that might lead to extinction.

The reader also might think about whether species *speciate* or *are speciated*; whether species can *interbreed* with one another; or whether the process of *reproductive isolation* is actually a process that involves species acting as discrete entities.

In swimming through the murky waters that are species concepts, we might ask what process is most critical with respect to species as entities (Cracraft 1987, 1989a). Framed another way, if we use a species concept to individuate entities, which process, or processes, do we expect those entities (species) to participate in? We ask these questions because if a definition individuates entities that actually participate, as entities, in no process (leaving aside whether this is a possibility in the first place), then what use is that definition? Relatively little, it would seem. And what does it imply about the "reality" of those "species"?

Most authors in this volume seem to agree—even if they do not state so outright—

that whole species do not speciate. Parts of species speciate in the sense that populations become isolated (at least under an allopatric model) and differentiate. At some point, and depending on the particular definition one adopts, that differentiated entity is called a species. In other words, species are speciated—they are the inferred end product of a process we call speciation. Given this, the reader should ask what kinds of entities would be expected to result from this type of process, and which definition(s) might be expected to individuate species/entities of this sort and which definition(s) might not. Let us frame the problem another way, and more directly: if species are only the end-product of a process of differentiation and participate in no other process as discrete entities, which species concept (or concepts) fits these entities and allows us to study them? Some do, and some do not. If a concept individuates things that are not a result of a process of differentiation, then that concept must exist for some other reason. This is part of the problem "how do we know what we think we know?" Such a concept might be individuating fictitious entities with regard to their participation in processes.

Another aspect of the problem "how do we know what we think we know?" involves *ancestral species* (or *stem species*). This is an issue that raises its ugly philosophical and empirical head in quite a few essays in this book, and some authors have sharp differences of opinion. They debate, for example, whether ancestral species can be said to maintain existence after a speciation event or whether ancestral species can be identified in the first place. The concept of ancestral species intersects with our ontological worldview (can we really say species are ancestors?) and with the epistemological tools we bring to bear on identifying taxa and determining their relationships, especially how we see and analyze character variation. It also intersects with the previous discussion. Presumably we are talking here about a discrete real entity, an ancestral species, and the process of *splitting*.

In some of these discussions of species concepts, ancestral species are spoken of as if one actually sees them (this language has been especially prevalent within paleontology). That, of course, is not what is happening. What one sees are individual organisms (actually, parts of them) in space and time, along with phenotypic variation among these specimens that is then interpreted to draw boundaries among populations and taxa. If these specimens extend into the contemporary world, perhaps we have other kinds of associated information (e.g., genetic, ecological, behavioral).

Drawing species limits among fossils entails many difficulties, and designating something as an ancestor entails even more. If in a discussion of ancestral species one does not refer to character information, but instead waxes and wanes about lineages, genetic continuity, branching events, and all other sorts of theory-laden language, then one is not dealing with the real world. Lineages and branching events are theoretical constructs; if they are part of a definition, then logically that definition says little about "how we know what we think we know." For that, one must individuate entities on the basis of characters and use that information, perhaps along with stratigraphy and geography, to make inferences about splitting. This is not, of course, an argument against using terms such as *lineage*, *genetic*, or *branching* in a definition, only a realization that the use of theoretical terms, in the absence of empirical evidence for their substantiation, may lead us astray.

Things to Look For

Some definitions of species operate within a frame of "how we think the world might be structured." That is, they define species in terms of the theoretical entities that might be supposed to participate in some evolutionary process of differentiation. Other concepts define species in terms of what we can recover (end products). And some concepts combine a little bit of both. They have different implications as far as "how we know what we think we know." And they also reflect a further debate in these essays between those systematists, on the one hand, who do not believe a definition has to be operational but only provide a theoretical framework (through a definition) of what the entities of nature are (species) and those systematists, on the other, who think a definition should point to a way of identifying species-level entities.

These represent two different ways of viewing how a species definition should be structured—more theoretical or more operational—but the distinction is not always clear-cut because some definitions straddle the extremes. What is worth looking for is what the systematist advocates beyond the definition. If the definition is heavy on theory but light on how species are to be identified in the real world, then one must look for the methodology that allows that definition to be applied. If the definition is very operational in its approach, then one must look to see whether the entities identified by that approach make theoretical sense with respect to some process-level phenomenon. In other words, somehow a species definition must be inclusive of an ontology and an epistemology. To have an ontological view of an entity that cannot be identified or studied is not much use, and to have methodological tools that cannot identify or allow us to study real entities is also not much use. The syntax of the definitions in this book, by and large, does not contain both aspects in equal amounts. That is why one must look to the discussion surrounding, supporting, and justifying that concept. To see how these concepts are applied in the real world is crucial; thus, one must go beyond these essays and delve into the literature. Once one does that, the distinctions among the concepts become sharper. In some instances, these definitions really do make a difference in how the world is described and studied. In other instances, real differences are few and far between, and the concepts are very close indeed to one another, even though the rhetoric may lead one to believe otherwise.

When reading the essays that follow, three things are worth keeping in mind. First, about 1.75 million species have been discovered and described. Second, with those

1.75 million species come two trenchant facts about our knowledge of this diversity: many of these species are known from a single specimen, and most are known from a limited amount of material. Moreover, the vast majority of the species come from a single point in space and time: populational variability is barely an issue, and these species have no fossil record. These generalities come to us because 90% or so of the world's diversity consists of insects and other invertebrate groups. And third, systematists estimate there are at least 8 million (and probably many more) species remaining to be discovered and described. No doubt the vast majority of these will also be characterized by few specimens from a single point in space and time.

These facts suggest that there is a gap between the real world as it is described by invertebrate diversity and the theoretical and philosophical literature on species concepts, which so often assumes the ideal in empirical evidence: populations with lots of individuals to allow assessment of variation, information on spatial distribution, perhaps some on their fossil history, and information about gene flow (interbreeding), ecology, behavior, and the like. This is a worldview that is largely based on vertebrates, yet even for these groups it often does not reflect the reality of the paucity of available data.

What does all this mean for species concepts? It implies that we should be careful in seeking justification for a particular species concept if it cannot embrace the vagaries of real-world data with aplomb. No hemming. No hawing. It must work. This does not mean that we should abandon theory and philosophy, ontology and epistemology, individuality, reality, pattern versus process, and all the other notions that orbit around discussions of species concepts. But we must keep our feet firmly planted on the ground.

PART 1

Position Papers (Point)

2

The Biological Species Concept

Ernst Mayr

I define biological species as groups of interbreeding natural populations that are reproductively isolated from other such groups. Alternatively, one can say that a biological species is a reproductively cohesive assemblage of populations. The emphasis of this definition is no longer on the degree of morphological difference, but rather on genetic relationship. This species concept represents a complete change in the ontological status of species taxa. For those who adopt the Biological Species Concept, species are no longer considered to be classes (natural kinds) that can be defined, but rather concrete particulars in the view of the biologist that can be described and delimited but not defined. Species status is the property of populations, not of individuals. A population does not lose its species status when an individual belonging to it makes a mistake and hybridizes. The word *interbreeding* indicates a propensity; a spatially or chronologically isolated population, of course, is not interbreeding with other populations but may have the propensity to do so when the extrinsic isolation in terminated.

The increasingly wide adoption of the Biological Species Concept was facilitated by the discovery of two fatal flaws in the Typological Species Concept. More and more often species were found in nature with numerous conspicuously different intraspecific phena—that is, differences caused by sex, age, season, and ordinary genetic variation—with the result that members of the same population sometimes differed more strikingly from each other than generally recognized good species. Conversely, in many groups of animals and plants, extremely similar and virtually indistinguishable cryptic species were discovered, the individuals of which, when coexisting, did not interbreed with each other but maintained the integrity of their respective gene pools. Such

This essay is largely based upon and incorporates extensive excerpts from Mayr (1996) by permission of the University of Chicago Press and Mayr (1988b) by permission of Kluwer Academic Publishers.

sibling species are perhaps more common in animals than in plants, but they certainly invalidate a species concept based entirely on degree of difference. Let us examine the origins of the Biological Species Concept.

Historical Considerations

The Biological Species Concept developed in the second half of the nineteenth century. Up to that time, from Plato and Aristotle until Linnaeus and early nineteenth century authors, one simply recognized *species, eide* (Plato), or *kinds* (Mills). Because neither the taxonomists nor the philosophers made a strict distinction between inanimate things and biological species, the species definitions they gave were rather variable and not very specific. The word *species* conveyed the idea of a class of objects, members of which shared certain defining properties. Its definition distinguished a species from all others. Such a class is constant, it does not change in time, and all deviations from the definition of the class are merely accidents, that is, imperfect manifestations of the essence (*eidos*). Mills in 1843 introduced the word *kind* for species [and Venn (1866) introduced *natural kind*], and philosophers have since used the term *natural kind* occasionally for species (as defined above), particularly after B. Russell and Quine had adopted it. However, if one reads a history of the term *natural kind* (Hacking 1991), one has the impression that no two authors meant quite the same thing by this term, nor did they clearly discriminate between a term for classes of inanimate objects and biological populations of organisms. There is some discussion among philosophers about whether there are several types of natural kinds, but I will refrain from entering that discussion. The traditional species concept going back to Plato's *eidos* is often referred to as the Typological Species Concept.

The current use of the term *species* for inanimate objects such as nuclear species or species of minerals reflects this classical concept. Up to the nineteenth century this also was the most practical species concept in biology. The naturalists were busy making an inventory of species in nature, and the method they used for the discrimination of species was the identification procedure of downward classification (Mayr 1982, 1992a). Species were recognized by their differences; they were kinds, and they were types. This concept was usually referred to as the Morphological or Typological Species Concept.

Even though this was virtually the universal concept of species, there were a number of prophetic spirits who, in their writings, foreshadowed a different species concept, later designated the Biological Species Concept. The first among these was perhaps Buffon (Sloan 1987), but a careful search through the natural history literature would probably yield quite a few similar statements. Darwin unquestionably had adopted a biological species concept in the 1830s in his *Transmutation Notebooks,* even though later he largely gave it up (Kottler 1978, Mayr 1992b). Throughout the nineteenth century quite a few authors proposed a species definition that was an approach to the Biological Species Concept (Mayr 1957a).

Late in the nineteenth century and in the first quarter of the twentieth century, taxonomists such as K. Jordan, E. Poulton, L. Plate, and E. Stresemann were among those who most clearly articulated the Biological Species Concept, as will be shown below.

As long as the inventory taking of kinds of organisms was the primary concern of the students of species, the Typological Species Concept was a reasonably satisfactory concept. But when species were studied more carefully, all sorts of properties were discovered that did not fit with a species concept that was strictly based on morphology. This was particularly true of behavioral and ecological properties. Most damaging was the discovery of the unreliability of morphological characters for the recognition of biological species.

Morphological difference had traditionally been the decisive criterion of species. Population A (e.g., continental North American savanna sparrows) was determined to be a different species from population B (e.g., savanna sparrows from Sable Island, Nova Scotia) if it was deemed to be sufficiently different from it by morphological characters. This definition was very useful in various clerical operations of the taxonomist such as in the cataloguing of species taxa and their arrangement in keys and in collections. However, for two reasons it was inadequate if not misleading for a study of species in nature. The first one is that, as is now realized, there are many good biological species that do not differ at all, or only slightly, morphologically. Such cryptic species have been designated *sibling species.* They occur at lesser or greater frequency in almost all groups of organisms (Mayr 1948). They are apparently particularly common among protozoans. Sonneborn (1975) eventually recognized 14 sibling species under what he had originally considered a single species, *Paramecium aurelia.* Many sibling species are genetically as different from each other as morphologically distinct species. A second reason a morphological species concept proved unsatisfactory is that there are often many different morphological types within a biological species, either because of individual genetic variation or different life history categories (males, females, or immatures), which are morphologically far more different from each other than are the corresponding morphological types in different species.

The morphological difference between two species fails to shed any light on the true biological significance of species, the Darwinian *why* question. So-called morphological species definitions are in principle merely operational instructions for the demarcation of species taxa. The realization of these deficiencies of the Typological Species Concept led, in due time, to its almost complete replacement among zoologists by the so-called Biological Species Concept.

Many of the authors who profess to adhere to the Morphological Species Concept do not seem to realize that unconsciously they base their decisions ultimately on the reproductive community principle of the Biological Species Concept. They combine drastically different phenotypes into a single species because they observe that they were produced by the same gene pool. This had already been done by Linnaeus when he synonymized the names he had given to the female mallard and the immature goshawk.

The biological significance of species is now clear. An organization of the diversity of life into species permits the protection of well-balanced, well-adapted gene pools. Numerous authors have arrived at this conclusion, and it was most recently confirmed by Paterson (1973:32) himself when he said that the study of speciation is "the study of the mechanisms by which isolating mechanisms, which protect the gene pool of a species from introgression, come into existence." There is only one other way by which superior gene combinations can be protected, and that is by a shift to uniparental reproduction (asexuality). The question that we posed at the beginning, as to the *why* of speciation, is now answered, and this answer represents the consensus of current evolutionary biology.

The next question to be answered, and it cannot be emphasized too strongly that this is an entirely independent question, is, by what devices is the integrity of a species being maintained? Dobzhansky (1935, 1937) introduced the term *isolating mechanisms* for these devices. He called them "physiological mechanisms making interbreeding [with nonconspecifics] difficult or impossible" (1935:349). In 1937 (p. 230) he defined as an isolating mechanism "any agent that hinders the interbreeding of groups of individuals," producing as an effect that "it diminishes or reduces to zero the frequency of the exchange of genes between the groups."

Dobzhansky was already aware of the independence of the isolating mechanisms from other characteristics of species. "The genetic factors responsible for the production of the isolating mechanisms appear to constitute rather a class by themselves. Thus, mechanisms preventing a free interbreeding may apparently develop in forms that are rather similar genotypically, and vice versa, genotypically more different forms may remain potentially interfertile" (1935:352). Modern studies seem to indicate that in some cases just a few genes may control effective reproductive isolation, whereas in other cases even a rather profound genetic restructuring of populations may not result in reproductive isolation. This is of course an expected manifestation of the incidental nature of the origin of isolating mechanisms and of the prevalence of mosaic evolution.

All authors who have written on isolating mechanisms—for instance, Dobzhansky (1937:228–258) and Mayr (1963:89–109)—have stressed the enormous diversity of such devices. In addition to sterility genes, chromosomal incompatibilities, and ecological exclusion, behavioral properties that facilitate the recognition of conspecifics are most important in higher animals. The existence of such behavioral mechanisms has been well known to naturalists, presumably far back into the nineteenth century or even earlier. Paterson (1978:369) quoted an excellent statement by W. Peterson. Plate (1914) articulated the recognition concept by stating that "the members of a species are tied together by the fact that they recognize each other as belonging together and reproduce only with each other." I stated that "species are a reproductive community. The individuals of a species of higher animals recognize each other as potential mates and seek each other for the purpose of reproduction" (Mayr 1957a:13). I have always stressed the importance of recognition and devoted several years to an experimental analysis of the sensory cues involved in the reproductive isolation between different species of *Drosophila* (Dobzhansky and Mayr 1944; Mayr and Dobzhansky 1945; Mayr 1946a, 1946b, 1950).

In view of the long-standing and widespread realization of the important role of recognition in the maintenance of the integrity of the species, it is curious that Paterson thought that he had invented an entirely "new concept of species, the Recognition Concept, which is conceptually quite distinct from the current paradigm, the Isolation Concept" (1980:330). In consequence, Paterson defined the species as the "most inclusive population of individual biparental organisms which have a common fertilization system" (1985:25). In all of his more recent publications, Paterson has stressed the great difference between his new Recognition Concept of species and the old Biological Species Concept. One is somewhat puzzled by this claim when one reads Dobzhansky's definition of the biological species: "the largest and most inclusive . . . reproductive community of sexual and cross-fertilizing individuals which share in a common gene pool" (1950:405). Even though virtually all modern evolutionists define a biological species as a reproductive community, Paterson (1981, 1985) has insisted that "the Biological Species Concept is essentially equivalent to the Isolation Concept." This claim is correct only insofar as the two concepts answer the question of the *why* of species. The Biological Species Concept, however, also answers the question about the *how* of species. Indeed, as Raubenheimer and Crowe (1987) correctly pointed out, the behavioral subset of the isolating mechanism specified by Mayr and by Dobzhansky in their supposed Isolation Concept is precisely the characteristic on which Paterson based his Recognition Concept. The fact that Paterson pleaded for the acceptance of his Recognition Concept from 1973 to 1986 in so many (at least six) publications indicates that he is rather disappointed that it has not found broader acceptance.

1. The so-called Recognition Concept does not specify, as does the Biological Concept, what the actual role of species is in nature. It answers the *how* but not the *why* question, as was pointed out above.

2. The term *recognition* is deeply flawed. Many authors such as Paterson, Plate, and numerous students of behavior have been aware of the recognitional aspects of such encounters. I referred to it in the following analysis: "The specific reaction of males and females toward each other is often referred to loosely as 'species recognition.' This term is somewhat misleading, since it implies consciousness, a higher level of brain function than is found in lower animals. . . . '[S]pecies recognition,' then, is simply the exchange of appropriate stimuli between male and female to ensure the mating of conspecific individuals and to prevent hybridization of individuals belonging to different

species" (Mayr 1963:95). This statement was followed by nine pages of discussion of such ethological stimuli classified on the basis of the principal sense organs involved.

3. Paterson's restriction of isolating mechanisms to behavioral recognition excludes all species with postmating isolating mechanisms. Furthermore, if the term *recognition* is rigidly construed, this species concept would virtually exclude all plant species. However, Paterson actually used the word *recognition* very broadly, comparing it with the "recognition of a specific antigen by its specific antibody" (1985:25). Consistent with such a broad definition, he also considered interactions "such as [those] between pollen and stigma, or between egg and spermatozoa," as recognition (1985:25). It is not quite clear why he did not also include here postmating isolating mechanisms effected by an interaction of incompatible chromosomes or genes.

Enlarging the concept of recognition, however, does not solve all problems. There is usually a considerable asymmetry between the sexes. Although females are usually highly discriminating, males in many, if not most, species have a rather generalized image of potential mating partners. Males of a given species in many genera of animals are ready at all times to mate with females of other congeneric species, or even with females of rather distant genera. Males would belong to different species than the females owing to Paterson's concept because they have a different recognition pattern.

The large number of new species concepts and species definitions proposed in recent years well reflects the seemingly utter confusion in this field. It seems to me that there are four reasons for this state of affairs.

1. We have experienced in the past 250 years the gradual, but only partial, replacement of the previously dominant Morphological Species Concept, based on typological essentialism, by a so-called Biological Species Concept, as discussed above. What the scientist actually encounters in nature are populations of organisms. There is a considerable range in the size of populations, ranging from the local deme to the species taxon. The local deme is the community of potentially interbreeding individuals at a locality (see also Mayr 1963:136), and the species taxon has been referred to by Dobzhansky as the "largest Mendelian population." The task of the biologist is to assign these populations to species. This requires two operations: (a) to develop a concept of what a species is, resulting in the definition of the species category in the Linnaean hierarchy, and (b) to apply this concept when combining populations into species taxa.

A number of recent writers on the species problem have failed to appreciate that the word *species* is applied to these two quite different entities in nature: species taxa and the concept of the category species. As a result, their so-called species definition is nothing but a recipe for the demarcation of species taxa. This is, for instance, true for most of the recent so-called phylogenetic species definitions. It is also largely true for Templeton's (1989, 1994) Cohesion Species Concept. A paper often cited as a decisive refutation of the Biological Species Concept (Sokal and Crovello 1970) is perhaps an extreme example of the confusion resulting from the failure to discriminate between the species as category (concept) and as taxon.

THE SPECIES TAXON

The word *taxon* refers to a concrete zoological or botanical object consisting of a classifiable population (or group of populations) of organisms. The house sparrow (*Passer domesticus*) and the potato (*Solanum tuberosum*) are species taxa. Species taxa are particulars, individuals, or biopopulations. Being particulars, they can be described and delimited against other species taxa.

THE SPECIES CATEGORY

Here the word *species* indicates a rank in the Linnaean hierarchy. The species category is the class that contains all taxa of species rank. It articulates the concept of the biological species and is defined by the species definition. The principal use of the species definition is to facilitate a decision on the ranking of species-level populations, that is, to answer the question about an isolated population: Is it a full species or a subspecies? The answer to this question has to be based on inference (the criteria on the basis of which such a decision is made are listed in the technical taxonomic literature, e.g., in Mayr and Ashlock 1991:100–105). A complication is produced by the fact that in the Linnaean hierarchy asexual species are also ranked in the species category, even though they do not represent the Biological Species Concept.

The literature traditionally has referred to the "species problem." However, it is now clear that there are two different sets of species problems, one being the problem of how to define the species (what species concept to adopt) and the other being how to apply this concept in the demarcation of species taxa. It is necessary to discuss these two sets of problems separately.

2. Some authors find the old typological concept more convenient in the delimitation of species taxa than the biological one (see the discussion of typological concepts above).

3. This new concept is not applicable to asexually reproducing organisms. The Biological Species Concept is based on the recognition of properties of populations. It depends on the fact of noninterbreeding with other populations. For this reason, the concept is not applicable to organisms that do not form sexual populations. The supporters of the Biological Species Concept therefore agree with their critics that the Biological Species Concept does not apply to asexual (uniparental) organisms. Their genotype does not require any protection because it is not threatened by destruction through outcrossing. Any endeavor to propose a species definition that is equally applicable to both sexually reproducing and asexual populations misses the basic characteristics of the biological species definition (the protection of harmonious gene pools).

4. The word *species* is applied both to the taxonomic category species (the rank in the Linnaean hierarchy) and to the species taxon. Surprisingly, many authors do not realize the different meaning of the word *species* in the two different contexts.

The criterion of species status in the case of the Morphological (Typological) Species Concept is the degree of phenotypic difference. According to this concept, a species is recognizable by an intrinsic difference reflected in its morphology, and it is this which makes one species clearly different from any and all other species. A species under this concept is a class recognizable by its defining characters. It is what philosophers call a *natural kind*. A museum or herbarium taxonomist who has to sort numerous collections in space and time and assign them to concrete and preferably clearly delimited taxa may find it more convenient to recognize strictly phenetic species in these cataloguing activities. I will presently point out the difficulties this causes.

Eventually it was realized by perceptive naturalists that species of organisms are not the same as the natural kinds of inanimate nature, and some biologists began to grope for a new species concept. However, they did not truly find it until Darwin had made it legitimate to ask *why* questions in biology. It was necessary to ask: Why are there species? Why do we not find in nature simply an unbroken continuum of similar or more widely diverging individuals? (Mayr 1988b). The reason, of course, is that each biological species is an assemblage of well-balanced, harmonious genotypes and that an indiscriminate interbreeding of individuals, no matter how different genetically, would lead to an immediate breakdown of these harmonious genotypes. The study of hybrids, with their reduced viability (at least in the F^2) and fertility, has demonstrated this. As a result, there was a high selective premium for the acquisition of mechanisms, now called isolating mechanisms, that would favor breeding with conspecific individuals and inhibit mating with nonconspecific individuals. This consideration provides the true meaning of species. The species is a device for the protection of harmonious, well-integrated genotypes. It is this insight on which the Biological Species Concept is based.

It is only incidental that such a species based on the concept of a reproductive community also has other properties, such as the occupation of ecological niches that are sufficiently different so as to provide competitive exclusion.

The Application of the Biological Species Concept

The Biological Species Concept is based on local situations where populations in reproductive condition are in contact with each other. The decision of which of these populations are to be considered species is not made on the basis of their degree of difference. They are assigned species status on a purely empirical basis, that is, on the observed criterion of presence or absence of interbreeding. It is the empirically determined interbreeding that is decisive, not the degree of difference. Observations in the local situation have clearly demonstrated the superior reliability of the interbreeding criterion over that of degree of difference. This conclusion is supported by numerous detailed analyses of local biota. I refer, for instance, to the plants of Concord Township (Mayr 1992a) and the birds of North America (Mayr and Short 1970). In particular, there is no difficulty when there is a continuity of populations and gene flow results in genotypic cohesion of the assemblage of populations.

Before going on to an analysis of more difficult situations, let me repeat that the Biological Species Concept is inapplicable to asexual organisms. They form clones, not populations. Because asexual organisms maintain their genotype from generation to generation by not interbreeding with other organisms, they are not in need of any devices (isolating mechanisms) to protect the integrity and harmony of their genotype. In this I entirely agree with Ghiselin (1974b).

Most of the criticisms of the Biological Species Concept are directed against the decisions made in applying the Biological Species Concept in the delimitation of species taxa. Using the Biological Species Concept as a yardstick in ordering contiguous interbreeding populations causes no difficulties. However, the criterion of interbreeding would seem to be inapplicable in the delimitation of species wherever isolated populations are involved, populations isolated either in time or space. I have presented in great detail the reasoning used by the defenders of the Biological Species Concept when assigning such populations to biological species (most recently in Mayr 1988a, 1988b, and 1992a). I will now summarize my arguments but refer to the cited publications for further detail.

The basic difficulty is that every isolated population is an independent gene pool and evolves independently of what is going on in the main body of the species to which it belongs. For this reason, every peripherally isolated population is potentially an incipient species. Careful analysis of their genetics and the nature of their isolating mechanisms has indeed shown that some of them are on the way to becoming a new species, and some of them actually have already passed this threshold. In areas suitable for such a distribution pattern, particularly in insular regions, every major species is usually surrounded by several populations that have reached the stage of being allospecies, but as far as all of them are concerned, we must make an inference on the basis of all available data and criteria as to how far along they have proceeded on the way to becoming a separate species. When making this inference, we must be clearly conscious of what we are actually doing. We are studying the available evidence (properties of species populations) in order to determine whether or not the species concept (the definition of the concept) is met by the respective populations. The logic of this procedure has been well stated by Simpson (1961:69; see also Mayr 1992a:230). This means that we do not combine two populations into one species because they are similar. Rather, we conclude that they are so similar because they belong to the same species. Molecular biology, of course, has given us far more evidence on which to base our conclusions than the purely morphological evidence previously available to a taxonomist. The greatest practical difficulty encountered by the investigator is the occurrence of mosaic evolution. Populations may acquire reproductive isolation but only minimal morphological difference (resulting in sibling species), whereas other populations may acquire conspicuously different morphologies but no isolating mechanisms. Equally, rates of molecular divergence and the acquisition of niche specializations vary independently of the acquisition of reproductive isolation.

Even accepting all these difficulties, it is evident that the endeavor to use all the available evidence to arrive at the correct decision may provide a far more meaningful classification biologically than an arbitrary decision simply to use degree of morphological difference. To be sure, assigning populations to biological species on the basis of the set of criteria discussed by Mayr (1969:181–187) will not eliminate the possibility of an occasional mistake. However, it is the best method available to a biologist.

Most ecologists and students of behavior study a given local biota. And it is quite irrelevant to them whether a distant, peripherally isolated population is called a subspecies or a species. The work of a student on the song sparrows of San Francisco Bay, for instance, is in no way affected by the decision on whether to treat the song sparrows of the Aleutian Islands as subspecies or as a separate species. The taxonomic rank of such isolated populations is of major concern only to the cataloguer and curator of collections. It must be emphasized in this connection that the difficulties of assigning geographic isolates do not in any way weaken the Biological Species Concept and its profound biological-evolutionary significance.

In most ambiguous situations it is, however, advantageous to treat allopatric populations of doubtful rank as subspecies: "the use of trinomials conveys two important pieces of information: (1) closest relationship and (2) allopatry. Such information is valuable, particularly in large genera. Geographical replacement suggests, furthermore, that either reproductive isolation or ecological compatibility has not yet been evolved" (Mayr and Ashlock 1991:105).

In ornithology the convention has developed in recent years to call strongly differentiated allopatric populations allospecies, indicating that these populations have reached the degree of morphological distinctness ordinarily characterized by full species. To call such populations subspecies or full species is irrelevant for most biological investigations (for a more detailed discussion as to how to carry out the inference of species status, see Mayr 1988b and Mayr and Ashlock 1991).

Criticism of the Biological Species Concept

Why is the Biological Species Concept, even though so widely adopted, still so often attacked? An analysis of numerous papers critical of the Biological Species Concept leads me to the conclusion that the criticism is almost invariably due to a failure of the critics to make a clear distinction between the species category (species concept) and the species taxon. The Biological Species Concept (and species definition) deal with the definition of the species category and the concept on which it is based. This concept, protection of a harmonious gene pool, is strictly biological and of course only has meaning where this gene pool comes into contact with the gene pools of other species, that is, at a given locality at a given time (the nondimensional situation). Only where two natural populations meet in space and time can it be determined what is responsible for the maintenance of their integrity. There is never any doubt in sexually reproducing species that it is the reproductive barrier. Two closely related sympatric species retain their distinction not because they are different in certain taxonomic characters, but because they are genetically programmed not to mix. The definition of monozygotic twins, as Simpson (1961) pointed out so rightly, provides a homologous causal sequence. Two similar brothers are not monozygotic twins because they are so similar, but they are so similar because they are monozygotic twins. It is the concept of reproductive isolation that provides the yardstick for delimitation of species taxa, and this can be studied directly only in the nondimensional situation. However, because species taxa have an extension in space and time, the species status of noncontiguous populations must be determined by inference.

Because I recently presented a detailed analysis of a number of criticisms of the Biological Species Concept, I will not repeat myself but simply refer to the analysis (Mayr 1992a:222–231). Here I will answer only a few criticisms that have been made more recently.

Learning that the Biological Species Concept reflects the nondimensional situation, Kimbel and Rak (1993:466) concluded that it is a "failure of the Biological Species Concept to explain the temporal persistence criterion of individuality." This objection confuses the species concept with the delimitation of species taxa. One arrives at the species concept under the condition of nondimensionality, but species taxa have, of course, an extension in time: they are not newly created in every generation. The Biological Species Concept presents us with the great advantage of providing a yardstick that permits us to infer which populations in space and time should be combined into one reproductively cohesive assemblage of populations and which others should be left out. As we shall presently see, none of the competing species concepts has such a criterion.

I want to emphasize particularly that *evolving* is not such a species criterion, as has been claimed by a number of recent authors. Species do not differ in this respect from other living entities. Of course, every species is a product of evolution, but so is every population, every isolate, every species group, and every monophyletic higher taxon.

Criticism of Competing Species Concepts

The reason that, despite the vigorous advocacy of several competing concepts, the Biological Species Concept continues to be so widely adopted is that adherents of the

Biological Species Concept feel that the other concepts have serious weaknesses. Although this is not the place for a detailed analysis of the competing concepts, including the phylogenetic one, a short review of these weaknesses is necessary in order to explain the continuing popularity of the Biological Species Concept. Among the competing concepts not treated in this book, the following may be considered.

1. The Nominalist Species Concept. According to this concept, a species is nothing but a subjective bracketing together of individuals or populations under a name. Every naturalist knows that species are not such arbitrary constructs. The species of birds in our gardens, woods, fields, and marshes are well-defined, real entities. Nothing convinced me of the fallacy of the nominalist species concept as much as the discovery that the primitive Stone Age natives of the mountains of New Guinea recognized essentially the same entities as species as a Western academically trained systematist did. Such a reality of species is documented for all sexually reproducing species. For additional comments on the nominalist species concept, see Mayr (1988a:317).

2. The Typological Species Concept. A typological species is an entity that differs from other species by constant diagnostic differences, but what one may consider a diagnostic difference is totally subjective. The so-called Typological Species Concept is actually not a concept at all, but simply an arbitrary measure for delimiting species taxa. The result of this procedure is classes (natural kinds) without the biological properties of species. Long before the time of Linnaeus, the typological species definition ran into considerable practical difficulties owing to striking morphological differences within a species and owing to the existence of morphologically similar or identical sibling species. I have published several detailed critiques of the Morphological Species Concept (Mayr 1963, 1969, 1988a:316, 1992a:223).

3. The Recognition Species Concept. Some recent authors have adopted H. Paterson's so-called Recognition Species Concept. Its weaknesses were exposed by Coyne et al. (1988) and by Raubenheimer and Crowe (1987). Paterson's concept is nothing but the Biological Species Concept under a different name. I have shown in a detailed analysis what misconceptions induced Paterson to think he had a new concept (Mayr 1988b; see also above). I myself had considered, but rejected, the recognition terminology (Mayr 1963:95) because the term *recognition* implies "consciousness, a higher level of brain function than is found in lower animals." The term *recognition* is, of course, even less applicable to the sterility factors in plants and to the chromosomal incompatibilities responsible for most cases of postzygotic isolation. Recognition and isolation are simply two sides of the same coin. There is no separate recognition concept.

4. The Ecological Species Concept. The so-called Ecological Species Concept (Van Valen 1976), based on the niche occupation of a species, is for two reasons not workable. In almost all the more widespread species are local populations that differ in their niche occupation. An ecological species definition would require that these populations be called different species, even though, on the basis of all other criteria,

it is obvious that they are not. More fatal for the Ecological Species Concept are the trophic species of cichlids (Meyer 1990), which differentiate within a single set of offspring from the same parents. Finally, there are the numerous cases (but none exhaustively analyzed) in which two sympatric species seem to occupy the same niche, in conflict with Gause's rule. All this evidence shows not only how many difficulties an ecological species concept faces, but also how unable it is to answer the Darwinian *why* question for the existence of species.

5. The Cohesion Species Concept. Perhaps Templeton's (1989, 1994) Cohesion Species Concept should be mentioned here. It attempts to combine the best components of several other species concepts, but fails to escape the resulting conflicts. It emphasizes the presence of gene flow, but fails to distinguish between the internal (isolating mechanisms) and external (geographic isolation) barriers to gene flow; it stresses cohesion through gene flow, but claims also to be applicable to taxa reproducing asexually, which have no gene flow. It attempts to characterize an evolutionary lineage, but does not indicate how to delimit such an open-ended lineage at either end; and Templeton does not state how to deal with the geographic variation of demographic-ecological attributes in widespread polytypic species. I do not see any advantages of this concept over the Biological Species Concept.

Pluralism

Looking at the current literature, one has the impression that different biologists prefer different species concepts. The Biological Species Concept has been almost universally adopted by students of behavior, by most ecologists (particularly those involved with the interaction of populations and species in nature) and those animal taxonomists who do generic and family revisions, as well as by the molecular biologists (Avise and Ball 1990).

Taxonomists who deal with scattered samples of taxa (most paleontologists), with the cataloguing of collections, with the cladistic ordering of higher taxa, with plant taxonomy (how widely the Biological Species Concept is accepted by botanists is being argued), and with asexual organisms prefer not to have a definite species concept but certainly have a methodology to delimit species taxa. It is my hope that the various groups, by studying my analysis, will acquire a deeper understanding of the Biological Species Concept and appreciate the reasons why they themselves prefer a different concept of species.

3

The Hennigian Species Concept

Rudolf Meier and Rainer Willmann

Hardly any concept in biology is as important or as controversial as the species concept. Yet, there is hardly a scientific paper in biology that does not at least implicitly use a species concept by generalizing the results of studies on individuals to the species level (Osche 1984:164; Sudhaus 1984:183). We will argue that a modified version of Hennig's species concept (Hennig 1950, 1966) is not only compatible with phylogenetic systematics, but suits the needs of biologists in other fields as well. We shall not enter the species-as-individuals discussion but offer our viewpoint that species as described by the modified Hennigian Species Concept are entities produced by nature.

Conceptual History

Hennig (1950) supported a species concept based on interbreeding that he had largely adopted from Naef (1919). However, Hennig realized that a species concept based on the criterion of "reproductive community" alone does not satisfy the demands of strict phylogenetic systematics because it cannot be applied to the temporal dimension of species. Any concept potentially useful in phylogenetic systematics must precisely specify the limits of species in time. Hennig proposed such a criterion when he argued: "When some of the tokogenetic relationships among the individuals of one species cease to exist, it disintegrates into two species and ceases to exist. It is the common stem species of the two daughter species"[1] (Hennig 1950:102). Thus, according to Hennig, stem species do not survive speciation events.

[1]"We call such relationships 'tokogenetic' that exist among individuals that are capable of producing offspring" (Hennig 1950:45–46).

During the first three decades following the publication of his species concept in "Grundzüge einer Theorie der phylogenetischen Systematik" in 1950, Hennig's concept gained little popularity, although he discussed it on numerous occasions (Hennig 1950, 1953, 1957, 1966, 1982, 1984). It conflicted with the prevalent concepts of that time, particularly with Mayr's Biological Species Concept (Mayr 1942, 1963) and Simpson's Evolutionary Species Concept (Simpson 1961). Furthermore, most neontologists were concerned with recent faunas only and were not particularly interested in the historical dimension of species, whereas many paleontologists were concerned with the search for missing links and ancestors and not with the consequences of splitting events. Hennig's proposal for delimiting species in time gained acceptance only among a few mainly German and Scandinavian zoologists (e.g., Brundin 1966; Bonde 1977, 1981; Griffiths 1974 [in Canada]; Königsmann 1975; Klausnitzer and Richter 1979; Ridley 1989; Richter and Meier 1994). Recently, it has been revitalized and modified by Willmann (1985a, 1986, 1991) and subsequently adopted by a number of zoologists and paleontologists (e.g., Ax 1987; Lauterbach 1992; Sudhaus and Rehfeld 1992). Hennig's species concept has been explicitly rejected by others (e.g., Peters 1970:19–20, 28; Mayr 1974:109–110; Wiley 1978:84–85, 1981a:34; Bell 1979; Hull 1979:432; Eldredge and Cracraft 1980:130–131).

Definition

Hennig proposed that all individuals connected through tokogenetic relationships constitute "a (potential) reproductive community and that such communities should be called species" (Hennig 1950:45–46). Here we would like to propose the following modified definition for the Hennigian Species Concept: "Species are reproductively isolated natural populations or groups of natural populations. They originate via the dissolution of the stem species in a speciation event and cease to exist either through extinction or speciation" (Willmann 1985a:80, 176; Willmann 1986).

Our definition differs from others such as Mayr's (1963:21), Paterson's (1985), and Templeton's (1989) in that it stresses reproductive isolation over internal cohesion through gene flow. Hennig used cohesion and isolation interchangeably because he was obviously mainly interested in the delimitation of species in time (1966:56–58).

For many authors, the notion of gene flow (cohesion) is at least as important as the notion of isolation. Others have stressed reproductive isolation (e.g., "The essence of the Biological Species Concept is discontinuity due to reproductive isolation" [Mayr 1957b:379]). Although the concepts of gene flow and reproductive isolation are closely related and sometimes viewed as two sides of the same coin, important differences exist. Reproductive cohesion is not only found within species but also and particularly within populations and demes. Nevertheless, no proponents of a cohesion species concept would consider such populations and demes to be species. We would argue that *cohesion* or *reproductive community* can be a criterion for a species definition (Ghiselin 1974a:537; Willmann 1985a:47) only when it is applied to the most inclusive taxon in which interbreeding occurs (the "most inclusive Mendelian population" [Dobzhansky 1970:357]). But in this case, it is really the reproductive gap between this taxon and its next of kin that is important in delineating species boundaries, and what appears to be a cohesion concept is in fact an isolation concept. It is the existence of a specific reproductive gap that defines the identity of a species and prevents the exchange of genetic information between itself and its sister species (Willmann 1991).

Conventional isolation species concepts are also flawed in that they stress isolation of a species not from its sister species but from any other species. However, it is the formation of a new reproductive gap between new sister species that defines a speciation event. This is why *species* (e.g., Mayr 1963:19; Coyne et al. 1988:190) and *speciation* (Willmann 1985a) are relational terms. In other words, a species is a species relative to its sister species. It follows logically that if species are defined in reference to a specific reproductive gap, they must cease to exist during a new speciation event when a new reproductive gap is formed.

It may be the case that isolation mechanisms usually evolve as by-products of processes that take place quite independently in separated incipient species (e.g., Paterson 1985; Templeton 1989:161–162). However, the essential requirement of speciation is that reproductive gaps create taxa that are hierarchically related to each other and between which genetic information cannot be exchanged. Criticizing isolation concepts for emphasizing isolation mechanisms, as has been done (e.g., Paterson 1985), is nonsensical.

Agamotaxa

Taxa consisting of uniparental organisms originate in a way similar to bisexual species, namely, via a splitting event (figure 3.1) (White 1978; Cole 1985; Willmann 1985a:67; Frost and Wright 1988). However, each organism is reproductively isolated from all other uniparental organisms, and each is a potential founder to its own hierarchically organized clade (figure 3.1). (Willmann [1985a], however, doubts that the term *reproductively isolated* applies to these cases.) In contrast to individuals of populations of biological species, clades cannot exchange any genetic information. Because the relationships between individuals of agamotaxa are dramatically different from the netlike relationships within bisexual populations (figure 3.1), we find it misleading to apply one term, be it *species* or any other, to uniparental and biparental taxa alike (for a similar point of view, see Hennig 1950:57–58).

A splitting event in which a bisexual species gives rise to an agamotaxon also produces a single bisexual species because a new sister taxon for the biparental species originates. Hence, there are two kinds of speciation: one that produces two

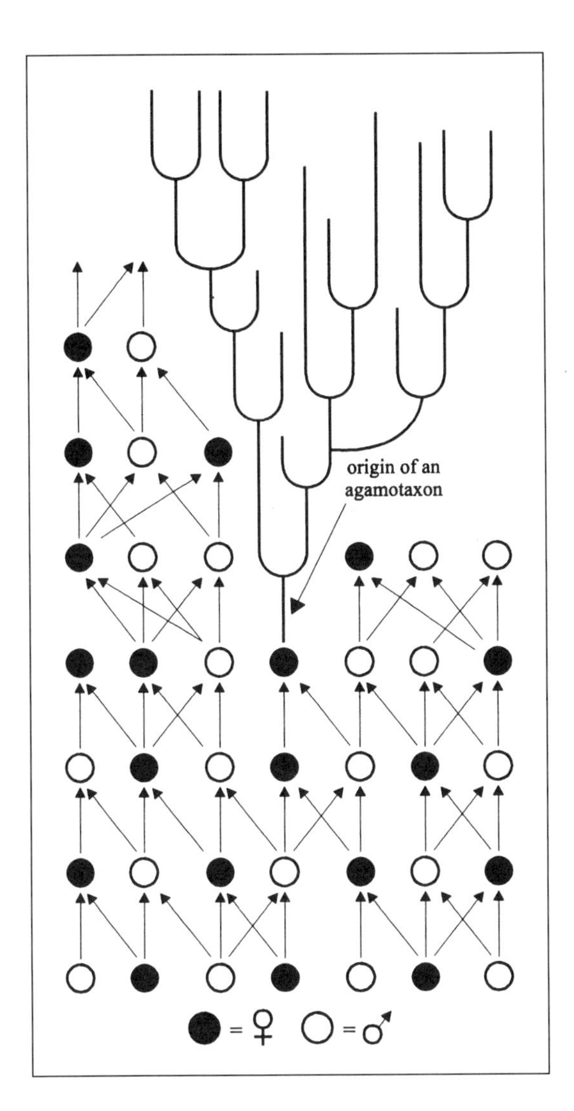

FIGURE 3.1.
The origin of an agamotaxon from a biparental species. Note the different organization within the biparental and uniparental taxa.

new (biparental) species and one that results in the generation of one species and one agamotaxon.

Despite the fact that uniparental taxa are dramatically different from bisexual species, taxonomists continue to describe uniparental "species," and there is even some debate over which taxa, if any, should receive specific recognition (e.g., Cole 1985; Walker 1986; Frost and Wright 1988). Descriptions of agamotaxa are usually based on overall similarity, although a voluminous body of literature exists that demonstrates the arbitrariness of phenetic techniques (e.g., Johnson 1970; Sokal and Crovello 1970:40; Doyen and Slobodchikoff 1974:240; Ridley 1986:39). It may here suffice to summarize the main objection to morphological species definitions:

1. Often morphology varies continuously, and it is thus impossible to objectively delimit taxa within the evolutionary continuum.
2. Even if there are diagnosable clusters of organisms in one time plane, these differences disappear as soon as evolutionary change is traced back along the time axis.
3. The position of species boundaries depends on what features are chosen for characterizing the morphological units.
4. Even if the characters have been well chosen, different clustering methods may yield different clusters of similar organisms; that is, the choice of clustering algorithm is necessarily arbitrary.

Sometimes there are indeed more or less distinct morphological gaps that allow the recognition of similar phenotypes within agamotaxa, as is the case in uniparental Bdelloidea (Rotatoria) (Holman 1987), some clones of whiptail lizards (*Cnemidophorus*) (Cole 1985), and mosses in the genus *Tortula* (Mishler and Brandon 1987:407). These phenotypes are discrete in today's time plane, but the gaps between them disappear as we proceed backward along the time axis. Any delimitation of such agamospecies in time would thus be arbitrary, and it is only a historical artifact that the intermediates are not known. In other cases the intermediates are recent and form continuous morphoclines. Attempts to force them into species may result in a chaotic taxonomy (e.g., some parts of *Cnemidophorus* [Walker 1986:428] and *Hieracium* [Maynard-Smith 1986:8]).

Due to the hierarchical structure of agamotaxa, phylogenetic analyses following Hennig's principle can be conducted using individual organisms as terminals. Traditionally, attempts to delimit agamospecies with phenetic techniques have nevertheless not been undertaken within a phylogenetic framework, and no distinction has been made between apomorphies and plesiomorphies. Thus, some species are delimited based on apomorphies but others exclusively on plesiomorphies. This practice clearly obscures the hierarchical structure that is present within agamotaxa. It is surprising that even phylogenetically inclined systematists have overlooked this problem of morphology-based species concepts in uniparental organisms (but see Frost and Wright 1988).

Phylogenetic Justification

Organismic evolution is the evolution of the species and its subgroups, and phylogenetic reconstruction is based on the assumption that evolution has occurred. Evolution and phylogenesis are historical processes. Hence, a species concept appropriate for phylogenetic systematics must consider the historical dimension of species.

When Hennig developed the idea that a monophyletic group consists of a stem species and all its descendants, he recognized that species limits had to be precisely defined. His proposal that species should be viewed as a temporal series of populations

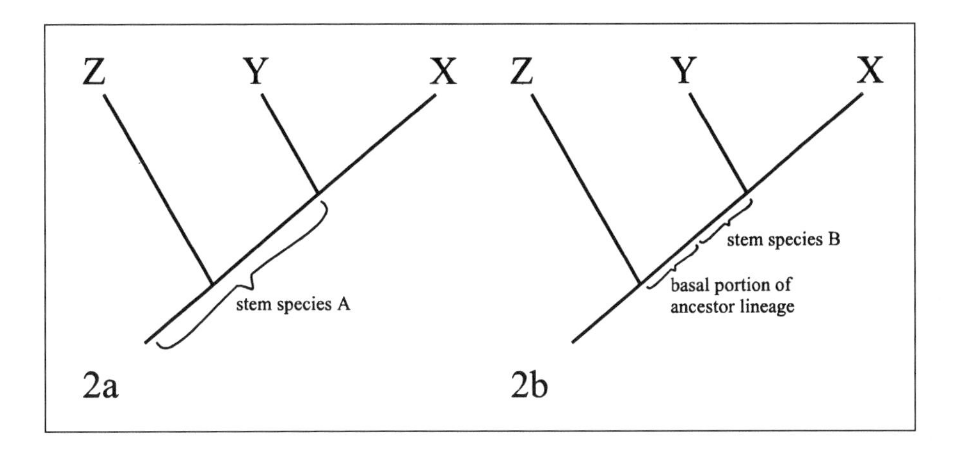

FIGURE 3.2.
(a) The survival of a stem species results in *X* and *Υ* not constituting a monophyletic taxon.
(b) Phyletic speciation.

connecting two speciation events follows logically from his definition of phylogenetic relationships. In 1957, he wrote, "a species *x* is only then more closely related to species *y* than to any other species *z* if it shares at least one common stem species with *y* that is not at the same time the stem species of *z*" (Hennig 1957:60). Imagine, in figure 3.2a, that the stem species A could "survive" and thus embrace more than exactly the segment between two speciation events. This concept has been proposed by numerous researchers (e.g., Wiley 1978, 1981a). In this case a monophyletic group consisting of *X* and *Υ* is no longer defined because there is no stem species shared only by these two species, as is required by the definition of monophyly. The survival of stem species thus leads to taxa consisting of closest relatives (sister taxa *X* and *Υ* in figure 3.2a) that no longer constitute a monophyletic group according to Hennig's definition of that term.

Now, consider the opposite case: delimiting a stem species that is smaller than a segment (*Hennigian species*; see figure 3.2b). The monophyletic group consisting of *X* and *Υ* can be defined as *X*, *Υ*, and stem species B in figure 3.2b, but in this case a basal portion of an ancestor lineage is left. The main argument against such a procedure is that the segment can only be arbitrarily divided into two portions. Furthermore, defining such species boundaries allows speciation without a splitting event (*phyletic speciation*), which most biologists would agree is not speciation at all because no novel evolutionary units originate from such an event (e.g., Wiley 1978:84, 1981a:34, 39, 41; Hull 1979:432; Eldredge and Cracraft 1980:114; Cracraft 1987:340; Ridley 1989:12; for a conflicting point of view, see Nixon and Wheeler 1990:219). If there is a need for a precise definition of *monophyly*, which there certainly is in phylogenetics, stem species cannot survive speciation and a species must comprise the entire branch segment between two speciation events.

Comparison with Other Concepts

We will briefly discuss the main differences between our species concept and the main competing concepts: the Phylogenetic, Evolutionary, Autapomorphic, and Biological Species Concepts.

THE PHYLOGENETIC SPECIES CONCEPT

According to Nixon and Wheeler (1990:218) a phylogenetic species is "the smallest aggregation of populations (sexual) or lineages (asexual) diagnosable by a unique combination of character states in comparable individuals (semaphoronts)" (see also Davis and Nixon 1992:427 and, for a related concept, Cracraft 1983, 1987, 1989a). Unlike the Hennigian Species Concept, this concept is based on morphology and thus suffers from the same flaws already discussed for morphological agamospecies.

Another objection to the Phylogenetic Species Concept is that its "diagnosable taxonomic units" can originate by the transformation of one phylogenetic "species" into another. Such *phyletic speciation* has nothing to do with the formation of new natural taxa (see also Hull 1979:432; Eldredge and Cracraft 1980:114; Cracraft 1987: 340). Nevertheless, phyletic "speciation" becomes rampant when the Phylogenetic Species Concept is applied.

Also, the application of the Phylogenetic Species Concept to asexual lineages allows the delimitation of species based entirely on plesiomorphies. After all, taxa displaying a combination of plesiomorphies are also "diagnosable by a unique combination of character states" (Nixon and Wheeler 1990:218). Considering that agamotaxa are organized hierarchically just like species groups, applying the Phylogenetic Species Concept to agamotaxa creates nonmonophyletic groups that obscure the phylogenetic structure within uniparental taxa.

THE EVOLUTIONARY SPECIES CONCEPT *SENSU* WILEY

According to Wiley (1978:80) a species is "a single lineage of ancestral descendant populations of organisms that maintains its identity from other such lineages and which has its own evolutionary tendencies and historical fate." This concept is entirely subjective (for a detailed critique, see Willmann 1989). There is no way to define "separate identities, tendencies and historical fates" of a "lineage" objectively, especially not when various degrees of hybridization are permissible, as Wiley has argued (1981a:27 et seq.). Rosen (1978:176) recognized that "the evolutionary species appears to conform, in practice, with Regan's (1926) definition that 'a species is what a competent taxonomist says it is.'"

Furthermore, in the Evolutionary Species Concept, the survival of the stem species is permitted, which leads to the problems outlined above: certain monophyletic groups are not defined because there are no stem species exclusive to those groups. Even more important in a biological context is that the Evolutionary Species Concept fails to recognize that *species* and *speciation* are relational terms.

THE AUTAPOMORPHIC OR PHYLOGENETIC SPECIES CONCEPT *SENSU* MISHLER AND THERIOT

"A species is the least inclusive taxon recognized in a classification into which organisms are grouped because of evidence of monophyly . . . that is ranked as a species because it is the smallest 'important' lineage deemed worthy of formal recognition, where 'important' refers to the action of those processes that are dominant in producing and maintaining lineages in a particular case" (Mishler and Brandon 1987:406; see also Rosen 1978; Donoghue 1985; Mishler 1985). The proponents distinguish between a grouping and a ranking criterion. We do not know what Mishler and Brandon deem worthy of recognition as a ranking criterion, but we are reasonably sure that taxonomists will adopt different criteria and that the choice of a ranking criterion is thus entirely subjective. Their grouping criterion is supposed to be *monophyly*. However, in taxa with netlike relationships among their members, derived character states do not necessarily indicate monophyletic groups, which do not even exist within tokogenies.

THE BIOLOGICAL SPECIES CONCEPT *SENSU* MAYR

According to Mayr (1942:120), biological species are "groups of actually or potentially interbreeding populations that are reproductively isolated from other such groups." Mayr (1963) stresses that the criterion of interbreeding is only applicable to a single time plane, whereas *biospecies* are historical entities. However, he failed to provide a criterion that specifies how and when biospecies originate and cease to exist (if not by extinction).

Another supporter of the Biological Species Concept, Bock (1979), insists that biological species are nondimensional with respect to time. He thus proposes a species concept that is not compatible with historical disciplines such as evolutionary biology and phylogenetics. Bock's concept cannot deal with either the origin or the evolution of species because his definition does not encompass the temporal dimension. Obviously, he fails to realize that everything that exists is necessarily historical (Willmann 1989:96).

Species Recognition

Questions concerning the nature of species and phylogenetic relationships immediately lead to questions about the recognition of natural entities and about the relationships among them. Such entities and relationships have certainly not come into existence with the purpose of being easily recognized, and one cannot therefore expect that they are necessarily easily recognizable.

Reproductive isolation is a real phenomenon in nature; it is isolation that keeps natural entities apart. Species concepts based on this criterion are therefore attempts to describe natural entities (Willmann 1991). If the criterion *reproductive isolation* is understood as absolute isolation, then the Biological Species Concept represents such an attempt, as does the Hennigian Species Concept. Willmann (e.g., 1985a, 1989, 1991) even argued that the Hennigian Species Concept is identical to the Biological Species Concept if absolute isolation is adopted as the criterion for contemporaneous populations and the origin of the isolation of two sister species is used to delineate species boundaries in time. In any case, it is important to stress that neither the Hennigian nor the Biological Species Concept is character related. Characters merely provide evidence as to where species boundaries are and can never be definitive species criteria. It is the detection of reproductive gaps that is decisive and that avoids arbitrary species boundaries and the "creation" of arbitrary species based on arbitrarily chosen sets of characters.

For the recognition of species in the same time plane, similar rules apply as for any population-based species concept. Phenetic evidence is used to distinguish populations that are likely to be isolated from each other, a task that is more easily accomplished for sympatric and parapatric populations than for allopatric ones. Davis and Nixon (1992: 430) claim that "if identical individuals can be drawn from two local populations (i.e., if no character distinguishes the two populations), the two populations belong to the same species." For us this conclusion is only a hypothesis.

Breeding experiments conducted in an artificial environment may be of use in determining whether allopatric populations belong to the same species (Eldredge and Cracraft 1980:98; Key 1981:439; Wiley 1981a:67–68; Sudhaus 1984:189). However, for many organisms such experiments cannot be performed, and the results are rarely conclusive because isolation mechanisms that may well exist in nature often break down under artificial conditions. When interbreeding produces fertile offspring, one may want to be conservative and consider the populations to be conspecific. If, however, individuals from allopatric populations that can be maintained and bred in the laboratory do not produce fertile offspring in crossing experiments, the populations probably belong to more than one species. The situation is somewhat different for plants. For many flowering plants their animal pollinators are an important component of their isolation mechanisms. Because breeding experiments in the laboratory are usually carried out without natural pollinators, isolation mechanisms commonly fail. This should not be interpreted as evidence that the populations are conspecific.

Sokal and Crovello (1970) and others have charged that species concepts based on the criterion of reproductive isolation are essentially phenetic. These authors confuse the species concept, which is based on species criteria, with characters that are used as evidence for species recognition (Mayr 1963). The recognition and description of a species result in a particular species hypothesis that may or may not be correct, that is, one that corresponds to a reproductively isolated unit (e.g., Hennig 1966:67;

Wiley 1978:79, 1981a:23; Eldredge and Cracraft 1980:94; Cracraft 1987:340; Willmann 1985a:60).

Character-related species concepts work well only if evolutionary change is not considered: every character starts in one or a few organisms of a population and becomes widespread only with time. In contrast, under the isolation concept, all available information is used for inferring species boundaries (e.g., including data about structure, color, behavior and physiology, ecology, etc.). We cannot expect the search for such boundaries to be a simple task. Probably one of the greatest merits of science is to face and explain difficult situations rather than avoid them.

Because a stem species may look the same as one of its two daughter species, the two may be distinguishable only by extrinsic evidence (such as age of the other daughter species). Under any character-related concept, however, the two would be lumped together, which does not solve the problem.

We thus agree that "the practical problem of recognizing species is clearly a secondary issue" (Ridley 1989:8). On this issue, we agree with Hull, who also pointed out that "some scientific terms, especially theoretical terms, are a good deal less operational than others; but, far from being regrettable, this situation is essential if theoretical terms are to fulfill their systematizing function and if scientific theories are to be capable of growth" (Hull 1968:438; for similar views, see Hull 1979:428; Eldredge and Cracraft 1980:94; Häuser 1987:245).

Potential and Actual Interbreeding

Some biological species definitions contain a reference to "potential" interbreeding (Mayr 1942, 1963), whereas others do not (Mayr 1969:26, 1970:12). Omitting the word *potential* from the definition has severe consequences. In particular, geographic separation then gains the same significance as reproductive isolation. If populations that have been separated from each other for only a short period of time are considered different species, then hybridization among such "species" becomes very common indeed. However, phylogenetic reconstructions are based on the assumption of hierarchical relationships among terminals and cannot adequately deal with nonhierarchical netlike relationships that result from such hybridization (see figure 3.1) (Brundin 1966:14; Hennig 1966; Nixon and Wheeler 1990:213; Wheeler and Nixon 1990:77; Davis and Nixon 1992:424). We maintain that for this reason, any definition of a species concept based on reproductive isolation that does not include the term *potential* will produce unstable "species" that may be related to each other in a netlike fashion when contact is reestablished and genetic information is exchanged. "Delimiting species on the basis of the potential to interbreed" is for us thus more than just "appealing in that it attempts to capture the idea that species exist through evolutionary time rather than being manifestations of current gene flow. Moreover, loss of the potential to interbreed guarantees that the entities are functioning as separate evolutionary units" (De Queiroz and Donoghue 1988:330; "point of no return," Willmann 1991:11). We wish to emphasize, however, that whether or not two individuals could potentially interbreed is not directly relevant to our species concept because it is not based on interbreeding, cohesion, or gene flow but on isolation only.

Discussion

REPRODUCTIVE ISOLATION

It cannot be overemphasized that absolute isolation is the only species criterion that excludes any arbitrariness (Key 1981; Willmann 1985a:46–47). Absolute isolation requires that even if hybrids between species occur, these hybrids are not able to successfully backcross with members of the parental populations. In the early twentieth century, geneticists and naturalists alike believed that reproductively isolated units closely corresponded to what had been considered species based on morphological criteria (e.g., Coyne et al. 1988:190). Later, however, it was found that the correspondence was not as perfect as had been anticipated (e.g., Ehrendorfer 1984). Especially among botanists, the Biological Species Concept was strongly criticized because many reproductively isolated units, sometimes referred to as *superspecies,* are highly polytypic. Many adherents of the Biological Species Concept departed from the strict use of the criterion of reproductive isolation and allowed for some degree of hybridization (Mayr 1957b, 1963; Ehrendorfer 1984:259; Coyne et al. 1988:196). This makes objective delimitations of species impossible. After all, the degree of "hybridization" that is considered acceptable is entirely arbitrary.

If some inbreeding is allowed between "species," morphology is no longer used exclusively as evidence for recognizing species, as it was in past practice, but is now used as an additional species criterion. The ambiguous usage of two different criteria at the same time creates problems. Ever since Aristotle it has been recognized that using two criteria in definitions leads to uninformative concepts because it conflicts with his *principio divisionis.* For example, classifying according to anagenesis and cladogenesis at the same time, as proposed by evolutionary systematists, results in uninformative systems. This flaw of evolutionary classification ultimately led to its decline. However, the same mistake is still being made by systematists whenever two species criteria, whether morphology, reproductive isolation, or any other (e.g., Doyen and Slobodchikoff 1974; Mishler and Brandon 1987:405 as "ranking criteria"), are used to delimit species (see De Queiroz and Donoghue 1990a:65). The problem is that one can never know which criterion was used in a particular case of species delimitation. Also, depending on which characters are considered in addition to reproductive isolation, different systematists will undoubtedly delimit "species" differently, and the task of describing species

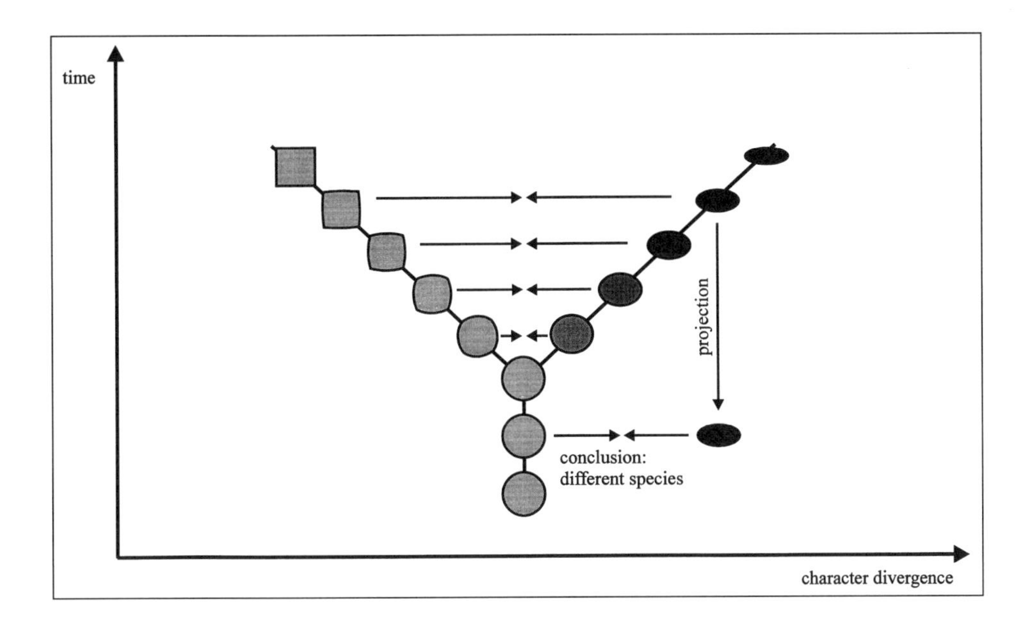

FIGURE 3.3.
Application of the isolation criterion to the time axis (for explanation, see text; after Willmann 1985a).

again reverts to an art instead of being science. Only when absolute isolation is used as the sole species criterion are objective and mutually comparable units delimited. Such objective boundaries of species are a prerequisite for counting species numbers and comparing biodiversity in different taxa.

The isolation criterion for species recognition does not interfere with any kind of biological research. One can do justice to morphologically different populations within species by describing them as subspecies. Moreover, using a strict isolation concept will be maximally efficient in pinpointing those evolutionarily interesting cases that reflect different stages of differentiation and possibly speciation.

Until recently, paleontologists who sympathized with the Biological Species Concept applied the isolation criterion not only to one time plane but also along the time axis. In doing so, they confused reproductive isolation with the continuous sequence of generations. Reproductive isolation is relevant only in one time plane because here it actually keeps gene pools apart that would otherwise merge. The same argument does not hold for generations that lived at different times. It is simply irrelevant and nonsensical to argue about whether two individuals that lived millions of years apart were reproductively isolated (Willmann 1985a:113–114). Furthermore, application of the isolation criterion to the time axis is subjective because the choice of a starting time plane is entirely subjective (figure 3.3).

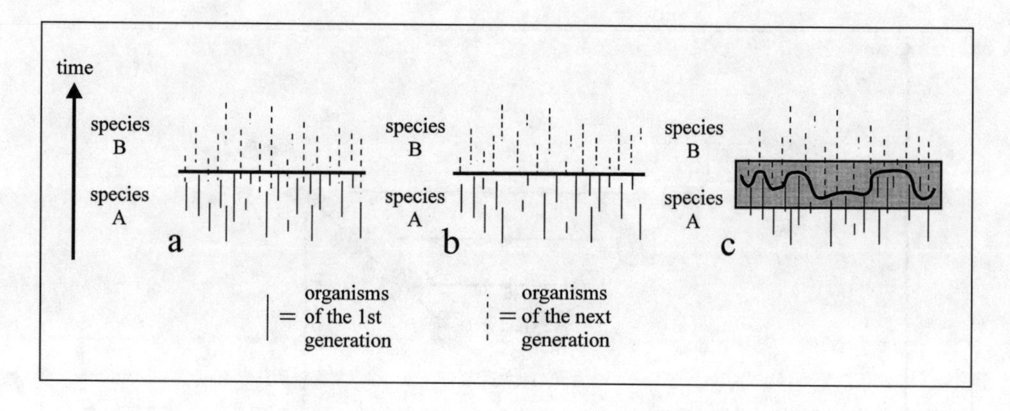

FIGURE 3.4.
Three different options for drawing species boundaries (after Willmann 1985a).

SURVIVAL OF THE STEM SPECIES

It is inherent to the Hennigian Species Concept that populations become species only relative to their next kin. Hence, speciation always creates a pair of new species, which in turn implies the dissolution of the stem species (Willmann 1985a:67, 1986, 1989: 108). Much of the criticism leveled against Hennig's position that the stem species does not survive the speciation event is based on *overall similarity* arguments. Frequently one of the daughter species is indistinguishable from its stem species, leading some authors to claim that it is not a different unit. This argument is based on a character-based species concept, whereas Hennig had adopted a concept similar to the biological one that is explicitly not based on similarity. Neither Hennig's nor Mayr's species definition makes any reference to morphology. Thus, criticizing their concepts from a phenetic point of view is beside the point. Also, it remains unclear why some of the same authors argue strongly in favor of recognizing sibling species but against the dissolution of stem species during speciation. For these authors, morphological similarity is judged unimportant in the case of sibling species, whereas it is considered a strong argument for the survival of the stem species.

The dissolution of the stem species in a speciation event has also been criticized because it draws the line between the stem species and its descendants between or even within a single generation (figure 3.4). However, this criticism overlooks the fact that all species concepts that delimit species in time face the same problem. Ever since the idea of spontaneous generation of new species was abandoned, all species are known to be connected through a continuous chain of generations. Accordingly, all species concepts necessarily divide this continuum somewhere within or between generations (Willmann 1985a:127). Criticizing such a procedure means denying that species have boundaries in time.

None of the authors who have criticized Hennig's criterion for delimiting species in time have been able to propose an alternative that allows the identification of objective temporal species boundaries. This was recognized by Platnick (1977a:97), who came to the conclusion that if species are individuals, "To attempt to divide a species between speciation events would indeed be arbitrary; . . . Dividing species at their branching points, however, becomes not only non-arbitrary but necessary."

4

The Phylogenetic Species Concept (*sensu* Mishler and Theriot): Monophyly, Apomorphy, and Phylogenetic Species Concepts

Brent D. Mishler and Edward C. Theriot

Conceptual History

Various attempts have been made at forging a species concept compatible with phylogenetic systematics or cladistics. Several such concepts have been called *the* Phylogenetic Species Concept, thus leading to considerable confusion in the literature. We support one version of the Phylogenetic Species Concept, one that, we will argue, can serve as a synthesis of all versions, but for historical clarity we will distinguish among different versions, their origins, and motivations (see also discussion by Baum 1992).

Hennig himself apparently held a view on species close to the Biological Species Concept. He defined species as "a complex of spatially distributed reproductive communities" (1966:47). He made an important distinction between tokogenetic relationships (ones that obtain between an "individual and its descendants and predecessors of the first degree" 1966:65) and phylogenetic relationships (ones that obtain between different lineages, "each bounded by two cleavage processes in the sequence of individuals that are connected by tokogenetic relations" 1966:20). In other words, tokogenetic relationships are diachronic, ancestor-descendant connections, whereas phylogenetic relationships are synchronic, sister-group connections.

Hennig's approach, although sound in many respects, errs in our opinion by postulating that there is one single breaking point at which reticulating tokogenetic relationship ends and divergent phylogenetic relationship begins. As we will discuss in detail below, there is not a clear cutoff point at which reticulation of lineages ceases, and furthermore, the point at which the possibility of reticulation goes to zero is well above the level at which cladistic structure can be reconstructed.

Hennig was also in error, in our opinion, when he used reproductive criteria to group organisms into species. The inappropriateness of using breeding compatibility in cladistic analysis was first pointed out by Rosen (1978, 1979) and Bremer and Wanntorp (1979). The fundamental problem is that the ability to interbreed (potential or actual), as a plesiomorphy by definition, is not a phylogenetically valid grouping criterion. Rosen argued (correctly, we think) that the basis for grouping in a cladistic system should be synapomorphy.

One phylogenetic species concept, based on unique patterns of shared characters, was proposed and defended by Eldredge and Cracraft (1980), Cracraft (1983), and Nixon and Wheeler (1990). For purposes of discussion, in this essay we shall refer to this general approach as the Phylogenetic Species Concept *sensu* Wheeler and Platnick, abbreviated Phylogenetic Species Concept (W-P). Oddly enough from a phylogenetic standpoint, unlike the concept of Rosen, this concept was explicitly not based on synapomorphy, but rather on a shared combination of characters (tracing back in this way to the species concept of Nelson and Platnick 1981). This approach to phylogenetic species followed the view of Hennig, seeing a fundamental break at the species level between diverging phylogenetic relationships above and reticulating tokogenetic relationships below.

Following up on some of Rosen's suggestions, Mishler and Donoghue (1982) presented and discussed another phylogenetic species concept (but they did not call it such until 1985 [Donoghue 1985; Mishler 1985]). For purposes of discussion we shall refer to this general approach as the Phylogenetic Species Concept *sensu* Mishler and Theriot, abbreviated Phylogenetic Species Concept (M-T). We are not particularly concerned about names for concepts; the concepts themselves are the important matter. However, we do not think priority is particularly important in such cases. Although we admit that Cracraft (1983) first used the name Phylogenetic Species Concept, we argue that a name should first and foremost reflect the meaning of a concept, and will point out in the next section that the Phylogenetic Species Concept (M-T) is uniquely suited for a classification based on phylogeny.

The empirical emphasis of Mishler and Donoghue (1982) was twofold. First, they pointed out the obvious noncorrespondence between groupings of organisms defined by different criteria. That is, an ecologically coherent group may be either less or more inclusive than the actively interbreeding group, and neither may correspond to a morphologically and/or genotypically coherent group. Second, as one looks at less inclusive and more inclusive groupings with respect to any one of these factors, there is no fundamental level, no level with some special reality for evolutionary studies. The theoretical emphasis of Mishler and Donoghue (1982) was also twofold. First, organisms should be grouped into species on the basis of evidence for monophyly, as at all taxonomic levels; breeding criteria in particular have no business being used for grouping purposes. Second, ranking criteria used to assign species rank to certain monophyletic groups must vary among different organisms, but might well include ecological criteria or the presence of breeding barriers in particular cases.

The two versions of the Phylogenetic Species Concept differ strongly in how they view reticulate relationships and characters in cladistic analysis. The Phylogenetic Species

Concept (W-P) argues that units having reticulate relationships are inappropriate for phylogenetic analysis (because that is inherently a study of branching relationships) and that such units can be the unambiguous phylogenetic species. "Fixed" combinations of characters were considered to be the empirical evidence for such units. The Phylogenetic Species Concept (M-T) argues that reticulation can occur throughout the hierarchy of life and so is not a special species problem, but rather one of more general difficulty. Under this view, apomorphies were considered to be the necessary empirical evidence for unambiguous phylogenetic species, as for phylogenetic taxa at all levels.

It might appear from the literature and the above discussion that the two basic versions of the Phylogenetic Species Concept are diametrically opposed. However, the differences can be overemphasized. Looking at the history of ideas and research groups in the manner pioneered by Hull (1988; see also Mishler 1987), it is clear that both in a phylogenetic sense and a phenetic sense, the two Phylogenetic Species Concepts are much closer to each other than either is to phenetic, biological, or evolutionary concepts. Both Phylogenetic Species Concepts have origins in the theory of phylogenetic systematics, and both emphasize that species be diagnosable. Differences in underlying philosophy remain, however. Wheeler and Platnick's Phylogenetic Species Concept has emphasized epistemology in its central focus on character evidence, whereas our Phylogenetic Species Concept has emphasized ontology in its central focus on monophyly. Difficulties in arriving at a synthesis of these two general phylogenetic approaches to species include finding the right balance between primary systematic patterns (i.e., character evidence) and evolutionary process theories. Clearly, it makes no sense to apply a species concept that requires prior specific knowledge of processes (e.g., reproductive behavior or ecological sorting). On the other hand, it is necessary that recognized species taxa be compatible with processes acting to produce phylogenies if phylogenetic classification is to be adopted as the general reference system. A unified Phylogenetic Species Concept can be proposed, based primarily on our Phylogenetic Species Concept in terms of its generalized ontological view about the meaning of phylogenetic criteria at any hierarchical level, but also incorporating the epistemological focus on character evidence from the Phylogenetic Species Concept (W-P).

Definition

The following paragraph provides a formal definition of our Phylogenetic Species Concept (based primarily on that of Mishler and Brandon 1987). The definition is complex, but then again so are the issues involved in producing hierarchical classifications from phylogenies:

> A species is the least inclusive taxon recognized in a formal phylogenetic classification. As with all hierarchical levels of taxa in such a classification, organisms are

> grouped into species because of evidence of monophyly. Taxa are ranked as species rather than at some higher level because they are the smallest monophyletic groups deemed worthy of formal recognition, because of the amount of support for their monophyly and/or because of their importance in biological processes operating on the lineage in question.

Some elaboration of terms from this definition is needed (see also Mishler and Brandon 1987). *Monophyly* is defined synchronically, following the "cut method" of Sober (1988), as all and only descendants of a common ancestor existing in any one slice in time. The ancestor is not an ancestral species, but rather a less inclusive entity such as an organism, kin group, or population that had spatiotemporal localization and cohesion/ integration (as discussed by Mishler and Brandon 1987). The synchronic approach is necessary to avoid the time paradoxes that arise when classifying ancestors with descendants (see discussion by Hennig 1966). Given that ontology, the evidence required for a hypothesis of monophyly is primarily corroborated patterns of synapomorphy (but may include other factors, such as geography).

The ranking decision (Mishler and Donoghue 1982; Donoghue 1985; Mishler 1985, Mishler and Brandon 1987) can involve practical criteria such as the amount of support for a putative group (e.g., number and "quality" of synapomorphies, bootstrap percentage, or decay index) and may also involve biological criteria in better-known organisms (e.g., the origin of a distinctive mating system at a particular node). This ranking decision is forced because systematists have legislatively constrained themselves to use a ranked Linnaean hierarchy. A larger issue is the recent call for reforming the Linnaean system by De Queiroz and Gauthier (1992) to remove the concept of ranks. Such a move would decrease the arbitrariness of ranking decisions at the species level as well, but the implications of this are beyond the scope of this paper (see Mishler 1999 for a discussion of the implications); we assume here that the current Linnaean system of ranked classifications is to remain in place.

Phylogenetic trees are the primary result of systematic study; they are hypotheses about nature, and thus "real" in that sense. However, any application of fixed names to phylogenetic trees (which result from continuous processes of divergence and reticulation) has to be arbitrary to some extent (particularly ranking). Grouping (based on monophyly) will be less arbitrary, but will still involve ancillary decisions about character homology and about how much support is necessary before one believes a hypothesis of monophyly. The main reason for providing a classification (beyond simply presenting the phylogeny, which could otherwise speak for itself) is to give a convenient handle, a name, for those monophyletic groups that we need to discuss or about which we need to record data. We need to name distinctive lineages as part of the process of inventorying, conserving, and using biological diversity. We also need to refer to specific phylogenetic groups in studies of processes acting to generate and maintain distinctive lineages. Not all discovered monophyletic groups, at whatever level, need to or should be named: some will be trivial in evolutionary terms (i.e., of short temporal duration or marked only by minor, selectively neutral apomorphies), some cryptic (i.e., marked only by molecular or chemical apomorphies and thus nearly impossible to distinguish for practical uses), and some poorly supported (and thus subject to frequent change as more characters and taxa are discovered). There will not always be a "smallest" monophyletic group in an ontological sense; monophyletic groups exist in many organisms (especially clonal ones, but also any group with limited dispersability) at much smaller levels than one would want to recognize formally with a Linnaean name. Thus, application of the species rank, like any other, is never automatic—it always requires independent justification.

Phylogenetic Justification

Our basic position is that there is no species problem per se in systematics. Rather, there is a taxon problem. Once one has decided what taxon names are to represent in general, then species taxa should be the same kinds of things, just the least inclusive. As discussed above, it must be recognized that there is an element of arbitrariness to the formal Linnaean nomenclatorial system. Evolution is real, as are organisms (physiological units), lineages (phylogenetic units), and demes (interbreeding units), for example. On the other hand, our classification systems are obviously human constructs, meant to serve certain purposes of our own: communication, data storage and retrieval, and predictivity. These purposes are best served by classification systems that reflect our best understanding of natural processes of evolution, and the field of systematics in general has settled on restricting the use of formal taxonomic names to represent phylogenetically natural, monophyletic groups. We will not repeat here the many reasons for preferring the phylogenetic approach to general-purpose classifications (see Hennig 1965, 1966; Nelson 1973; Wiley 1981a; Farris 1983); instead, having accepted principles of phylogenetic classification, we will argue for the thoughtful application of these principles to the species level.

A phylogenetic systematic study of a previously unknown group of organisms involves three major temporal, logical phases. To understand the uniquely phylogenetic basis for our approach to species, it is necessary to elaborate on these phases:

1. In the precladistic phase the elements of a cladistic data matrix are assembled. These elements include OTUs (operational taxonomic units), characters, and character states. OTUs are assembled initially from grouping together of individual specimens that are homogeneous for the characters then known (see also discussion by Vrana and Wheeler 1992). Hennig (1966) himself laid this process out quite well. In his words, "the individual is to be regarded as the lowest taxonomic group category" (1966:65). In Hennig's system, the individual organism is regarded as being composed of semaphoronts (character bearers), which are basically "the individual in a

certain, theoretically infinitely small, time span of its life, during which it can be considered unchangeable" (1966:65). Semaphoronts are connected by ontogenetic relationships to form the individual organism; individual organisms are connected by tokogenetic relationships to form ancestor-descendant lineages. Hennig was quite explicit (1966:66–70) in showing that although the above ontology is clear, the empirical process of grouping individual organisms together into hypotheses of species is far from clear. This complex process involves considerable reciprocal illumination (because developing hypotheses of distinct, independent characters with discrete states goes hand in hand with developing hypotheses of homogeneous OTUs). There is no "magic bullet," no obvious, theory-free way to individuate species. The process must involve analysis, and that analysis must be explicitly phylogenetic.

2. Cladistic analysis involves translation of the data matrix into a cladogram. Reciprocal illumination is often involved here as well because incongruence between characters or odd behavior of particular OTUs may lead to a return to phase 1, a reexamination of OTUs and characters, primarily to check for fit to the assumptions of the cladistic method (i.e., that OTUs should be homogeneous for the characters used and should be the result of a diverging phylogenetic process rather than a reticulating, tokogenetic process; characters should be discrete, heritable, and independent).

3. Classifications based on an assessment of the relative support for different clades provide a basis for evolutionary studies. Formal taxa (including species) are named here on the basis of clear support for their existence as monophyletic cross sections of a lineage and for their utility in developing and discussing process theories.

Discussion and Conclusions

Reticulation

Certain fundamental assumptions must be made in order to justify the use of cladistic parsimony for phylogenetic reconstruction. These have been discussed by a number of people (see summary by Sober 1988); Mishler (1994) argued that five basic assumptions are necessary:

1. Replication (in the sense of Hull 1980; Brandon 1990) must occur to form *lineages* (the diachronic ancestor-descendant relationship; Wilson 1995).
2. Particular features to be used as historical markers (*characters*) must have discrete variants (*character states* empirically, *transformational homologs* ontologically) that show a strong correlation (heritability in a population genetic sense) between parent and offspring.
3. Divergence (branching of lineages) must occur, as compared with reticulation, giving rise to patterns of *taxic homologs* (in the sense of Patterson 1982) shared among *sister groups* (the synchronic monophyly relationship).

4. Independence must occur among different characters; that is, no process (e.g., natural selection, gene conversion, developmental constraints) is operating to produce nonhomologous character associations that overwhelm taxic homologs, indicating common history.
5. Transformation in particular characters must occur at a relatively low rate, as compared with divergence (see Mishler 1994 for discussion and further literature references).

Note that the first and third assumptions are ontological, whereas the second, fourth, and fifth assumptions are merely epistemological. If one of the latter are violated to some extent, we can still get the true relationships. If the third assumption is violated by reticulation, true relationships of the resulting hybrid literally cannot be obtained via cladistic parsimony. Note that this is, of course, the case with any other phylogenetic reconstruction algorithm introduced, whether based on distances, phenetics, maximum likelihood, or some other criterion. However, there is hope for future development of algorithms to detect reticulation because it is possible to infer hybridity based on genomic studies (using chromosomal markers or allelic markers such as allozymes or RAPDs [randomly amplified polymorphic DNAs]; Rieseberg et al. 1990; Arnold et al. 1991; Rieseberg 1991; Arnold et al. 1992).

Reticulation is thus the *bête noire* for cladistics, as initially recognized by Hennig. There are a number of different sources of homoplasy (incongruency between certain character distributions and the cladogram based on maximum parsimony), such as adaptive convergence, gene conversion, developmental constraints, mistaken coding, and reticulation. The last-named factor is the most problematical because it involves the fundamental model of reality underlying cladistic analysis. The other factors are cases of mistaken hypotheses of homology, whereas homoplastic character distributions due to reticulate evolution involve true homologies whose mode of transmission is not treelike.

Hennig and later Nixon and Wheeler were correct in focusing on reticulation and the problems it causes for cladistics. Our opinion of the significance of this problem for the species question differs to some extent from theirs, however, for the following reasons: (1) just as barriers to reticulation are often not complete, reticulation is not a complete barrier to cladistic analysis; and (2) reticulate relationships range from intense (in panmictic, sexually reproducing groups, where individual relationships are exclusively reticulate) to less intense (in spatially or temporally subdivided groups, where both reticulate and divergent relationships exist, facultatively and/or obligatorily, among individuals).

The presence of some reticulation is not an absolute barrier to cladistic reconstructions. We can reconstruct relationships in the face of some amount of reticulation (how much is not yet clear, but is amenable to study). For example, McDade (1992) has shown that incorporating a few known hybrids in an analysis of "good" species does not seriously affect the cladistic topology of the good species. Of course, the hybrids cannot be placed correctly in a reticulate position solely via cladistic analysis, but the relationships of the nonhybrids may be perfectly reconstructable. McDade actually gives rules predicting what a hybrid taxon should do in a cladistic analysis; thus, there may be a self-correcting mechanism here, as there is with other sources of homoplasy; even major convergence (e.g., in cave animals) can be uncovered via cladistic analysis. As with convergence, where the application of cladistic analysis provides the only rigorous basis we have for identifying homoplasy and thus demonstrating nonparsimonious evolution (Farris 1983), the only way we can identify reticulation on the basis of character analysis alone is through the application of cladistic parsimony, followed by the examination of homoplasy to attempt to discover its source.

Furthermore, there is no consistently clear demarcation between reticulate and branching relationships. Hybridization takes place between clades of various patristic/cladistic degrees of relatedness. There is no sharp distinction between sexually versus asexually reproducing populations in a great many organisms. Bacteria exchange genetic material in a variety of ways. Diatoms, cladocerans, and rotifers commonly undergo many asexual generations, with occasional sexual generations occurring in response to environmental change; some lineages within these groups can be obligately asexual. In many diatoms, only part of a single clonal lineage can become sexual at any given time. Other forms of reticulation occur throughout nature. Rare, high-level hybridizations may occur among very divergent lineages, such as among genera of orchids; viral-mediated lateral transfer of genetic material is suspected at much higher levels.

Thus, just as there may be no largest cladistic unit for which reticulation is impossible, there may be no smallest irreducible cladistic unit within which no further diverging phylogenetic patterns occur; ontologically speaking, we are dealing with a fractal pattern. When one looks at a lineage closely, one sees a pattern of divergence of lineages within (and some reticulation, perhaps increasingly greater, as one looks at less inclusive lineages). Asexuals are the most extreme case; cladistic structure will go down to the organism level. This fractal pattern of reticulation and branching is a severe problem for phylogenetic inference by any means. But as argued above, phenomena such as symbiosis are discovered as incongruence between organismal and character phylogenies. Massive convergence in one character system is discovered by incongruence between that system and other characters. By presuming that synapomorphy is equivalent to strict taxic homology of sister groups, cladistic analysis implies that homoplasy is explainable by all other processes, including reticulation. Lacking other information, reticulation must always be presumed to be a possible explanation for homoplasy.

Assuming we want to discover reticulation by objective means (Vrana and Wheeler 1992), it will be important to focus further attention on the problem of reticulation. Were cladistic analysis to be attempted on individuals within a panmictic group, consensus cladograms would presumably be nearly completely unresolved. This would be the correct result: there is little or no cladistic structure to reconstruct in such cases. Admittedly, however, one might still get a single most parsimonious tree even with heavily reticulating units. An unproven assumption in such cases of intense reticulation among OTUs is that there would be a disproportionate number of nearly most parsimonious trees. One might also expect to observe nonrandom distributions of homoplastic characters (concerted homoplasy) in cases of hybridization. How modes of reticulation actually affect character distributions on cladograms is a productive avenue for empirical and theoretical investigations.

This avenue reflects one of the great strengths of the direct character analysis procedure of cladistics. Methods that sum information across all characters (distance or phenetic methods) instead of treating them discretely cannot directly discover reticulation. Although direct observation of reticulation (e.g., field studies of hybridization) would indicate that cladistic analysis is inappropriate for phylogenetic inference, the presence of fixed characters at some level of grouping is neither direct nor indirect evidence for reticulation below that level. Only homoplasy may be used as indirect evidence for reticulation.

In conclusion, reticulation is not a species-specific problem. Modes of reticulation may differ and may be more or less intense in different kinds of organisms. The central difficulty remains identifying reticulation events in the midst of cladistic events. At higher levels, there seems to be wide consensus that synapomorphy can be discovered in spite of reticulation. Our Phylogenetic Species Concept, a species concept that identifies species as taxa identifiable by apomorphy, is consistent with the entire phylogenetic system and in principle is no more or less vulnerable to violation of its assumptions than is any level of phylogenetic analysis.

ASEXUAL REPRODUCTION

Our Phylogenetic Species Concept as defined above is clearly equally applicable to sexual and asexual organisms. This is important because many lineages exist that reproduce solely or mainly by nonsexual means. On the other hand, despite claims to the contrary, Wheeler and Platnick's Phylogenetic Species Concept is not appropriate for asexual species, in part because it lacks clearly defined ranking criteria. Cladistic relationships exist down to the individual level in asexual species. Furthermore, plesiomorphically defined groups may be clades, but they are also likely to be simply grades or even polyphyletic assemblages, as is the case for higher taxa. Thus, only apomorphic characteristics can identify phylogenetically natural groups in asexual species; the only applicable concept here is the Phylogenetic Species Concept (M-T).

The species situation in clonal organisms was explored in detail in a series of papers in *Systematic Botany*, introduced by Mishler and Budd (1990). First of all, despite the impression given by certain writers in the field, there is no sharp distinction between sexually and asexually reproducing organisms (as discussed above). Every degree of frequency of sex exists among populations of different species, ranging from absolute asexuality, through rare fertilization events, to panmixia. One instance of sexual recombination in a million asexual generations does not suddenly change the ontological or epistemological status of a species. Secondly, the supposed difference in phylogenetic patterns between sexually and asexually reproducing organisms does not hold up under close examination.

Mishler (1990) addressed previous predictions about the discreteness of sexual versus asexual species, using a cladistic analysis of the moss genus *Tortula* (a clade within which a spectrum of sexuality occurs, ranging from frequent sexual reproduction to total asexuality). It would be predicted under standard evolutionary theory that sexual species should be more variable than asexual species within populations (because of recombination) and less variable between populations (because of the homogenizing effect of gene flow). Therefore, species in asexual groups should be less discrete than those in sexual groups. However, measures of species distinctness, either cladistic (i.e., the number of autapomorphies) or phenetic (i.e., ordinations or analyses of variance of morphometric data), showed no particular correlation with mode of reproduction. Mishler concluded that processes other than gene flow may be responsible for species formation and maintenance even in sexual groups, a finding that has implications for speciation studies (see Budd and Mishler 1990; Mishler and Budd 1990 for further discussion).

SPECIATION AND THE PHYLOGENETIC SPECIES CONCEPT

Theriot (1992) investigated patterns of speciation in relation to species concepts in a species complex of diatoms with an extremely robust fossil record. He took "phenotypically irreducible clusters" (i.e., groupings of organisms not divisible by cladistically significant characters; basically the Phylogenetic Species Concept of Cracraft 1983) as OTUs in a cladistic analysis and compared the resulting phylogeny with known ecological, stratigraphic, and biogeographic data. He concluded that three autapomorphic species each were products of evolution and probably also units participating in the evolutionary process, whereas the widespread, plesiomorphic *Stephanodiscus niagarae* is neither a product nor a unit of evolution. Thus, he cautioned against accepting the smallest phenetically recognized clusters of organisms as basic units or products of evolution.

A number of potential empirical errors can occur in analyses of species, including those conducted under our Phylogenetic Species Concept. However, there is one potential "error" (i.e., characters undiscovered at the time of analysis) for which our concept is robust with respect to other phylogenetic concepts. The diatom example again illustrates this point. Zechman et al. (1994) have begun to analyze these diatoms with molecular and morphological data, identifying cladistic structure within *S. niagarae*, further demonstrating its paraphyletic nature. An important point to be made is that even if cladistic structure could be demonstrated as real within the autapomorphic species, their interpretation as an evolutionary lineage would not be altered. However, the discovery of cladistic structure within the plesiomorphic species *S. niagarae* fundamentally shifts the view of *S. niagarae* as a natural unit to merely an aggregate of lineages. Thus, with regard to the primary goal of cladistic analysis and phylogenetic systematics, the discovery of natural groups, our Phylogenetic Species Concept applies a robust interpretation (i.e., that the identified group is monophyletic) to the discovery of new characters, whereas a concept lacking the use of apomorphy does not and cannot.

In general, our Phylogenetic Species Concept remains faithful to cladistic principles and thus is subject to exactly the same promise and problems of cladistic analysis that occur at any level. Any cladistic analysis that fails to take into account the possibility of reticulation may not be realistic. Not all lineages may have evolved apomorphic characteristics, and so they may not be identifiable through character analysis. That is, there may be monophyletic groups for which there is no direct evidence. Once again, this is a general problem for cladistic analysis and is not special to the species problem. On the other hand, if the standard assumptions of cladistic analysis are met, then our Phylogenetic Species Concept identifies natural units regardless of relationships among individuals of that unit.

5

The Phylogenetic Species Concept (*sensu* Wheeler and Platnick)

Quentin D. Wheeler and Norman I. Platnick

Conceptual History

Our Phylogenetic Species Concept has its origin in the writings of Willi Hennig (1966) and subsequent transformations of phylogenetic theory. Hennig recognized that the Biological Species Concept (Mayr 1942, 1963, 1969) was problematic relative to the chronological history of species and proposed modifications designed to fix this species concept (Hennig 1966; Ridley 1989). As phylogeneticists divorced the discovery of historical patterns of cladistic relationships from unnecessary assumptions about evolutionary processes (Platnick 1979), it became apparent that modes of speciation need not be confounded with criteria used to distinguish among species. This recognition and increased awareness of problems associated with the Biological Species Concept (e.g., Mishler and Donoghue 1982; Cracraft 1983, 1989a; Donoghue 1985; Mishler 1985) led to the formulation of several species concepts that were phylogenetic in name or substance.

Rosen (1978, 1979) first applied the principles of cladistic analysis to the problem of species recognition. The result was an autapomorphic concept of species that sought to delineate as species those groups of individuals (populations) sharing a unique apomorphy (see also Hill and Crane 1982). This autapomorphic species, "a geographically constrained group of individuals with some unique apomorphous characters, is the unit of evolutionary significance" (Rosen 1978:176). Subsequently, De Queiroz and Donoghue (1988) argued that species should be based on *monophyly,* seeking to do cladistic analyses below the level of species in order to arrive at groupings of populations based on supposed *synapomorphy.*

Independently, and simultaneously, Eldredge and Cracraft (1980) and Nelson and Platnick (1981) formulated a concept of species that was founded in phylogenetic theory yet independent of cladistic analysis: "a diagnosable cluster of individuals within which there is a parental pattern of ancestry and descent, beyond which there is not, and which exhibits a pattern of phylogenetic ancestry and descent among units of like kind" (Eldredge and Cracraft 1980:92) or "simply the smallest detected samples of self-perpetuating organisms that have unique sets of characters" (Nelson and Platnick 1981:12).

Cracraft subsequently restated his Phylogenetic Species Concept, deleting explicit mention of reproductive disjunction: "the smallest diagnosable cluster of individual organisms within which there is a parental pattern of ancestry and descent" (1983:170). This deletion is justified, in Cracraft's own words, because "I do not know a single example in which data on reproductive cohesion or disjunction are the sole factors establishing taxonomic limits. Indeed, even within sibling species, phenotypic differences of some kind, e.g., behavioral or biochemical, are always the primary data that lead to their recognition as distinct taxa" (1983:164). Additional problems with disjunction may be noted. Rosen (1979) observed that interbreeding is symplesiomorphic—that is, it was shared by individuals of the most recent common ancestor of two species—and that, therefore, its discovery among organisms in closely related species is neither surprising nor particularly informative.

The Phylogenetic Species Concepts advocated by Eldredge and Cracraft, Nelson and Platnick, and Cracraft all contain substantially similar components that were amplified further by Nixon and Wheeler (1990:218): "the smallest aggregation of populations (sexual) or lineages (asexual) diagnosable by a unique combination of character states in comparable individuals (semaphoronts)." This emerging consensus on a cladistic conception of species has several important, desirable properties and consequences. These will be explored following a formal restatement of the concept below.

The history of the Phylogenetic Species Concept progressed from efforts to fix the Biological Species Concept, to applying cladistic analysis to the problem, to formulating a concept fully compatible with phylogenetic theory but not dependent on prior cladistic analysis. All efforts at an alternative, "phylogenetic" species concept do not share this conceptual lineage. Two of these outlying approaches represent polarized logical extremes. One permits virtually any species concept to be applied "where appropriate," and the other allows for none of them.

Mishler and Donoghue (1982) suggested that because organisms vary so much, no single concept can apply equally well to all of them. Were this the case, the problem of finding an applicable concept would simply be replaced with an even more difficult one. How does one determine which of many possible concepts applies to a particular kind of organism? Because such pluralism introduces subjectivity into the process of applying species concepts, it becomes possible to make up ad hoc stories to account for any difficulties. If this pessimistic vision were true, one might prefer a process of adopting alternative species concepts and rejecting them one at a time. Once all possible concepts had been tested and found wanting, the free-for-all of pluralism might be adopted as a last resort. Unless potentially universally applicable concepts are examined individually, how can we hope to determine whether such a unifying concept exists? Despite the protracted controversy over species concepts, systematists have always worked under the assumption that a theoretically and practically sufficient species concept was attainable. Given the rapid progress since the publication of Hennig's work, this would appear to be a particularly unfortunate time to give up.

Vrana and Wheeler (1992:67) took the opposite extreme view, advocating that "individual organisms rather than any interbreeding 'group' in which they are placed should be used as the terminal entities in phylogenetic analysis." Those authors objected to the notion of a "taxonomic 'line of death' below which systematic analysis should not be attempted." We agree that the question of whether samples of a given set of populations show cladistic structure is ultimately an empirical one, not to be determined in advance by a prior decision that the populations constitute only one species. No one has ever suggested that the entities that happen to be called species in some existing classification are necessarily true individual species, rather than conglomerations of two or more species that we have not yet succeeded in diagnosing. And no one has ever suggested that work aiming to discover and rectify such errors "should not be attempted."

We nevertheless disagree with the conclusions of Vrana and Wheeler. There is every reason to expect that a sample of individuals taken from a single species, when analyzed cladistically, will show some particular set of relationships. There is no reason to expect that this particular set of relationships can be replicated when different traits are examined. For example, a morphological data set on adult spiders, with individual organisms taken as terminal taxa, would probably lead to all males clustering together with each other before any of them cluster with any females, whereas a data set of DNA sequences taken from the same individuals would presumably yield very different results. The fact is that a cladogram can be obtained by analyzing any single data set whatever; the only evidence that the relationships shown in the cladogram are not simply an artifact of the method is our ability to replicate the results—from a different data set for the same organisms. If the relationships among the terminal taxa are actually reticulate rather than hierarchical, there is no reason to expect that replicability to occur. That difference alone produces a theoretical "line of death" for cladistic analyses (figure 5.1), even if the question of where that line lies, in any particular case, must be determined empirically rather than by theory.

Furthermore, Hennig's (1966) distinction between tokogeny and phylogeny remains. Where tokogenetic relationships can be demonstrated (by empirical, testable observation), it is reasonable to assume that the requirements for either hierarchical relations among characters or phylogenetic relations among individuals (or populations) are not fulfilled. Interbreeding events, in this circumstance, result in reticulate patterns; offspring arise from two parents, not a unique beginner; and genes may shift in their commonality among individuals at any time and in any direction (cf. hierarchy *sensu* Woodger 1937). What, then, is the basis for assuming hierarchical structure

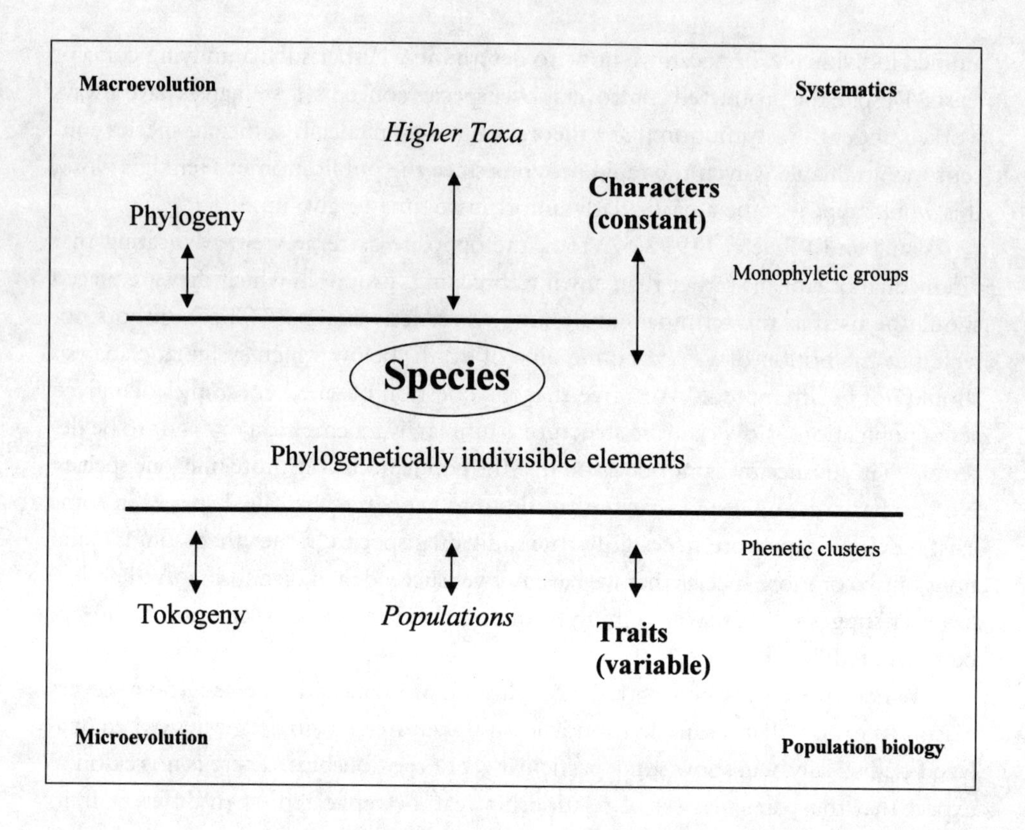

FIGURE 5.1.
Diagram depicting the line between tokogeny and phylogeny that corresponds to the species. Below this line, populations are characterized by variable traits. Above this line, species and monophyletic clades are characterized by constantly distributed characters. (Redrawn from Nixon and Wheeler 1990 by F. Fawcett.)

in tokogenetic data? And in the absence of hierarchy, on what basis are we to assume that a resulting cladogram may be interpreted as showing phylogenetic relations (Nelson and Platnick 1981)?

Definition of the Phylogenetic Species Concept

We define species as the smallest aggregation of (sexual) populations or (asexual) lineages diagnosable by a unique combination of character states. This concept represents a unit species concept. Phylogenetic species are at once the basic units of formal scientific nomenclature, Linnaean classification, and organic evolution. Under diverse evolutionary processes, ranging from clonal asexual forms to sexually reproductive organisms, phylogenetic species represent end-products. Without a concept that applies equally to all conceivable speciation processes, how could we hope either to measure biological diversity or to study the mechanisms causing speciation?

Our Phylogenetic Species Concept does share certain things held in common by species concepts in general (Nelson and Platnick 1981:11). For example, species may not be studied as wholes, making it necessary to study only samples of them. "The most we can say is that we have not yet been able to differentiate species within the sample" that we hypothesize to be a single species (Nelson and Platnick 1981:11). As a consequence, it is important that species concepts function as hypotheses that are themselves open to critical testing.

Like other species concepts, ours has an underlying assumption about self-perpetuation. This concept aims, like others before it, to recognize the kinds of life that perpetuate more of like kind. The concept would lose its unit-species benefits were it to fail to account for differences among individuals, between sexes, and among life stages. As Hennig (1966) observed, actual comparisons of species and of higher taxa are based on observations of organisms at particular life stages (semaphoronts).

One aspect of our definition vulnerable to criticism is its mention of populations that are, as will be discussed below, very difficult to define or recognize with precision. This, however, is a problem shared by every species concept that applies to sexually reproductive organisms. It is the legacy of a population research community that has concerned itself almost exclusively with processes, and the result of an assemblage of organisms that is inherently complex and unstable.

It is worth mentioning that when we initially distinguish species, we do so in preparation for and prior to a cladistic analysis. Therefore, the polarity of the characters used to recognize species need not be known with certainty; any hypotheses about polarity proffered prior to a rigorous application of the parsimony criterion would be tentative at best. Thus, for the purposes of distinguishing among species, reference is made to character states without regard to polarity. Although the correspondence between characters, homologs, and apomorphies is of critical importance to cladistics (Patterson 1982), such distinctions are irrelevant to species recognition. In this respect, our Phylogenetic Species Concept differs from that of Nelson (1989a), which treats species as identical to nonterminal taxa in all respects (most particularly as constituting relationships rather than groups), and therefore requires identification of apomorphies to diagnose species. Despite its very different metaphysics and ontology, in practice Nelson's view is therefore equivalent to the Autapomorphic or Monophyletic Species Concept.

Speciation is marked by character transformation. In turn, character transformation occurs through the "extinction" of ancestral polymorphism (see Nixon and Wheeler 1992a). The moment of speciation is, in theory, precise and corresponds to the death of the last individual that maintained polymorphism within a population. It is the removal of polymorphism through extinction that fixes a new species for a character state and that results in the transformation of the ancestral (polymorphic) attribute. Consider a simple example where ancestral polymorphism gives way to two daughter species, within each of which an alternative state for the character is fixed. Neither resultant state in this circumstance is more "apomorphic" than the other relative to the polymorphic mother population (figure 5.2). Thus, species need not be characterized by apomorphy (as, for example, the ancestral species of a clade). Assertions of De Queiroz and Donoghue (1988) notwithstanding, the concept of monophyly is simply inapplicable to species (Wheeler and Nixon 1990).

It has been suggested that our Phylogenetic Species Concept is problematic because it may result in an enormous number of species. What is the relationship, measured in numbers of species, between the Biological Species Concept and our Phylogenetic Species Concept? As Nelson and Platnick (1981) suggested, many well-substantiated subspecies will no doubt be elevated to species status. And we would predict that, in general, the Biological Species Concept has underestimated the number of end-products of evolutionary history by permitting subjective decisions about the inclusivity of polytypic species. Barrowclaugh, Cracraft, and Zink (personal communication) examined birds with this question in mind. They suggest that the Biological Species Concept has underestimated the number of species of birds significantly and that the actual number of living bird species is closer to 18,000 than to 10,000. Discrepancies with a similar level of magnitude were reported from an empirical study of the birds-of-paradise (Cracraft 1992).

Such dramatic increases, however, may be exceptional. A more anecdotal review of a dozen or so monographs of Coleoptera by one of us (Q.D.W.) suggests that the impact on beetles may be much less dramatic. Perhaps, however, less well-studied groups, particularly those for which subspecies have not been generally recognized, will show a closer parallel between the number of "biological" and phylogenetic species. Indeed, for spiders, one of us (N.I.P.) estimates that the species number is unlikely to be raised by even 1% because few arachnologists have ever used the Biological Species Concept to justify ignoring any of the taxa they could diagnose. The same is probably true of Coleoptera, except for a small number of taxa in which the Biological Species Concept has been used to justify combinations of consistently diagnosable population(s) (e.g., Barr 1979).

The issue of numbers of species is perhaps most threatening when asexual organisms are considered. One may well ask whether it is advisable to recognize every clonal form characterized by a mutation. It might well result in a great explosion of names for asexual organisms. Our response is, so what? If the goal of distinguishing species is to thereby recognize the end-products of evolution, should we seek to suppress naming large numbers of species where large numbers of differentiated end-products exist? How are we to learn whether one mode of speciation tends to produce greater or lesser species diversity than another mode of speciation unless we apply our species concept in a consistent and uniform way? It may be interesting to learn whether asexual reproduction produces more species than sexual reproduction, if that is in fact the case. Such questions cannot even be asked in the absence of a unit species concept and its rigorous application, regardless of the result.

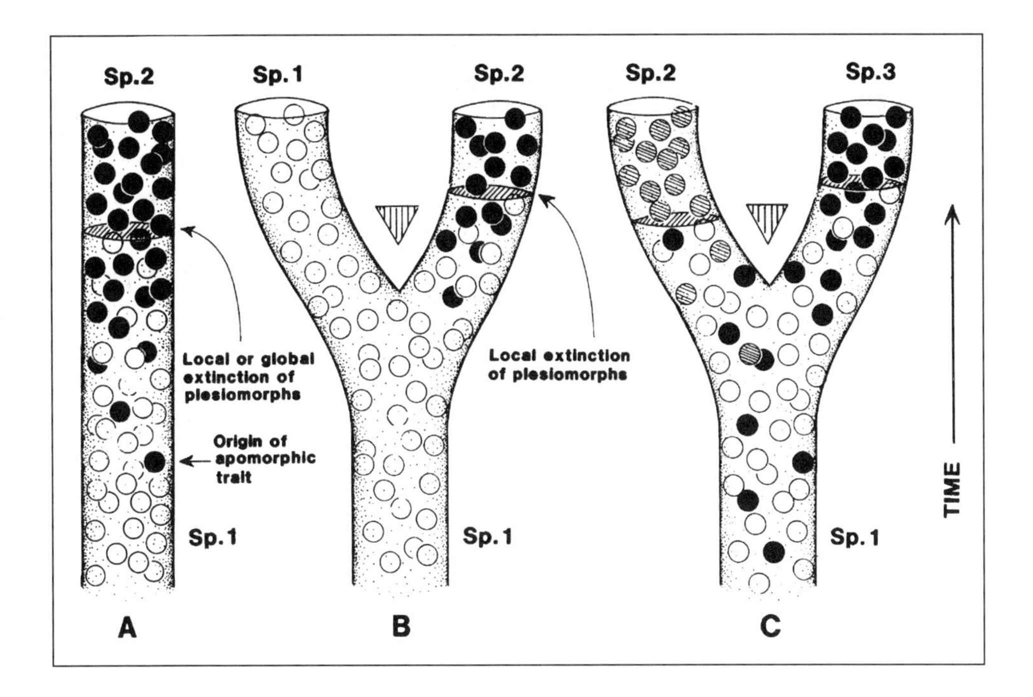

FIGURE 5.2.
The origin of species and character transformation are inseparably linked events given a phylogenetic species concept. Extinction of ancestral polymorphism results in the constant distribution of characters within daughter species. Examples of the origin of phylogenetic species include *anagenetic* (A) and *cladogenetic,* resulting in one (B) or two (C) new species. (After Nixon and Wheeler 1992a.)

Phylogenetic Justification

Pattern and Process

> Now what, one might ask, are processes of evolution? Do they not all presuppose the existence of a nonrandom pattern such as the [cladistic] one we have considered? No patterns—in general, no processes. No patterns, nothing to explain by invoking one or another concept of process. In short, a process is that which is the cause of a pattern. No more, no less" (Nelson and Platnick 1981:35). "By accepting the reality of previously recognized taxa, concepts associated with important biological processes are relegated to the role of after-the-fact explanations for the existence of these taxa, instead of functioning as central tenets from which real entities and the methods for their discovery are deduced. (De Queiroz and Donoghue 1988:318)

The parsimony criterion avoids unnecessarily complex assumptions about character transformations (Farris 1982, 1983). This, and a general effort to divorce the analysis of phylogenetic patterns from unnecessary assumptions about evolutionary processes, effectively transformed cladistics into a rigorous science (Nelson and Platnick 1981).

Our Phylogenetic Species Concept has the advantage of being insulated from specific assumptions about evolutionary processes. This, combined with the fact that the concept provides the elements for cladistic analysis, creates the basis for discovering the details of the patterns that hypotheses of evolutionary mechanisms claim to explain. This conceptual independence makes our Phylogenetic Species Concept compatible with virtually any credible speciation process imaginable (figure 5.3). Only processes at odds with the fidelic character distributions represented by species or monophyletic groups may be dismissed.

For a general discussion of the importance of divorcing pattern from process in phylogenetics, see Eldredge and Cracraft (1980) and Nelson and Platnick (1981). "The process of evolution" has become an unfortunate cliché in biology textbooks, perpetuating a sloppiness in the use of the word *evolution* that contributes to semantic confusion. Textbook definitions about changes in gene frequency notwithstanding, there is no "process of evolution." Instead, evolutionary patterns are the cumulative result of countless kinds of processes acting singly and in combination. Genetic drift, sexual selection, allopatry, allochronic sympatry, giant astroblemes on the earth's surface, and just plain old luck can all be evolutionary processes. Fortunately, the net result of evolutionary processes has been a single pattern of common ancestry, made retrievable by the hierarchical relationships among species' characters and the unique combination of character states distinguishing (phylogenetic) species.

TRAITS AND CHARACTERS, TOKOGENY AND PHYLOGENY

Hennig's (1966) distinction between tokogeny and phylogeny is illustrated clearly in figure 5.4. In the tokogenetic system, birth relationships are shared by individuals. The resultant pattern, due to panmixis, is reticulate rather than hierarchical. Even though a cladistic analysis of populations or individuals could be conducted and a branching diagram retrieved, such a "hierarchy" would be artifactual. The assumption of hierarchy is unwarranted where tokogeny has been observed. Parsimony has been shown to be the foundation for cladistic analysis (Farris 1982, 1983), but one may expect to detect an artifactual pattern when phylogeny is sought where hierarchical structure in data logically does not exist.

Tokogeny and phylogeny also differ in the nature of observable attributes. In the tokogenetic system, similarities are in the form of *traits*; in the phylogenetic system, synapomorphies are in the form of (transformed) *characters.* We use the terms *trait* and *character* in the restricted way suggested by Nixon and Wheeler (1990:217): "We consider all inherited attributes of organisms to be either traits or characters. Because of the interchangeability and confusion of these terms in the literature, it is necessary to distinguish between attributes that are not universally distributed among comparable individuals within a terminal lineage (traits) and those that are found in all comparable individuals in a terminal lineage (characters)."

Constancy of characters does not mean that the characters in question do not

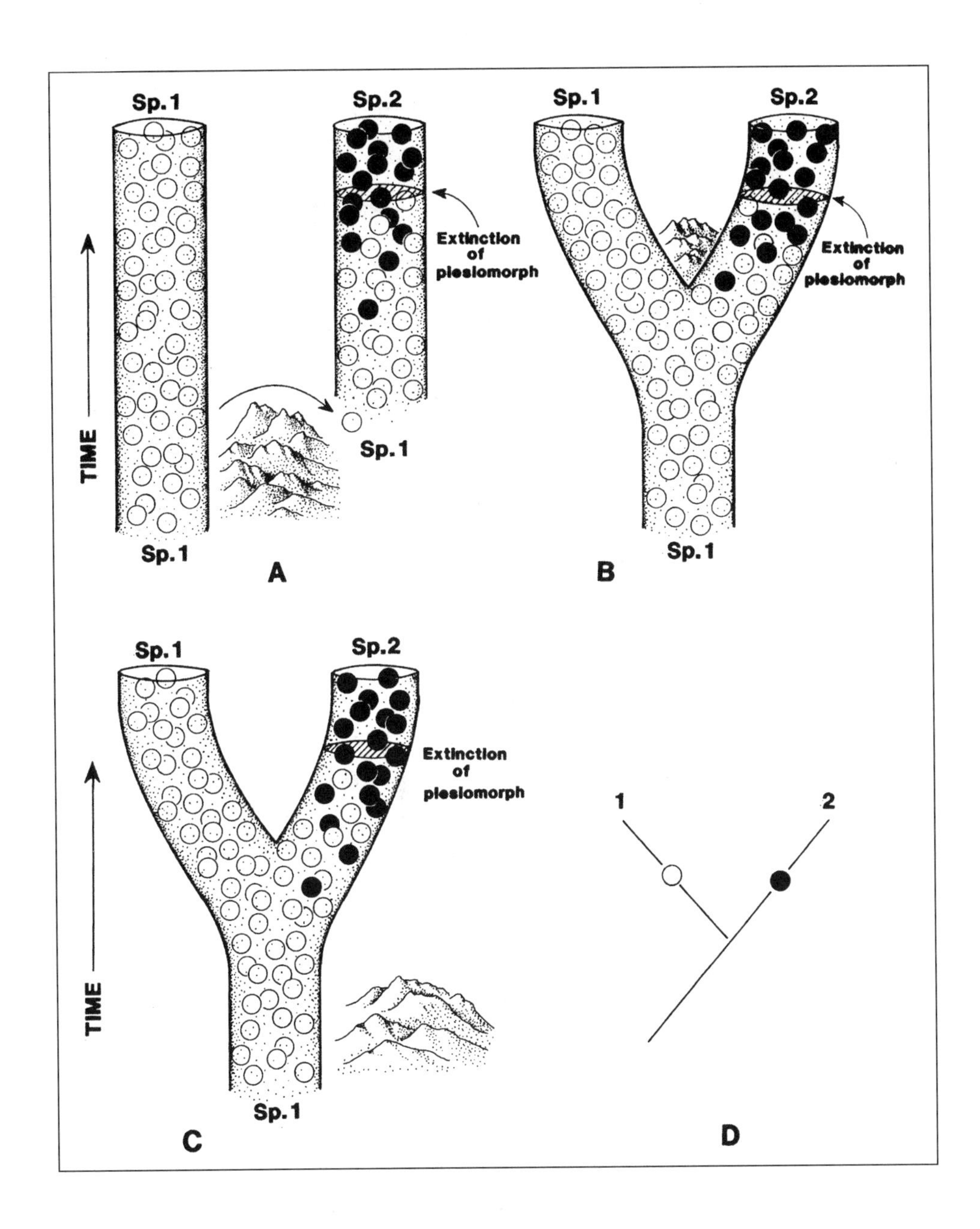

FIGURE 5.3.
Phylogenetic species are recognized by unique combinations of constantly distributed characters (D), regardless of the evolutionary process or processes responsible for speciation. For example, constant characters making species 1 and 2 diagnosable could result from dispersal and subsequent character transformation (A), vicariant isolation of two populations with subsequent character transformation (B), or sympatric speciation (C).

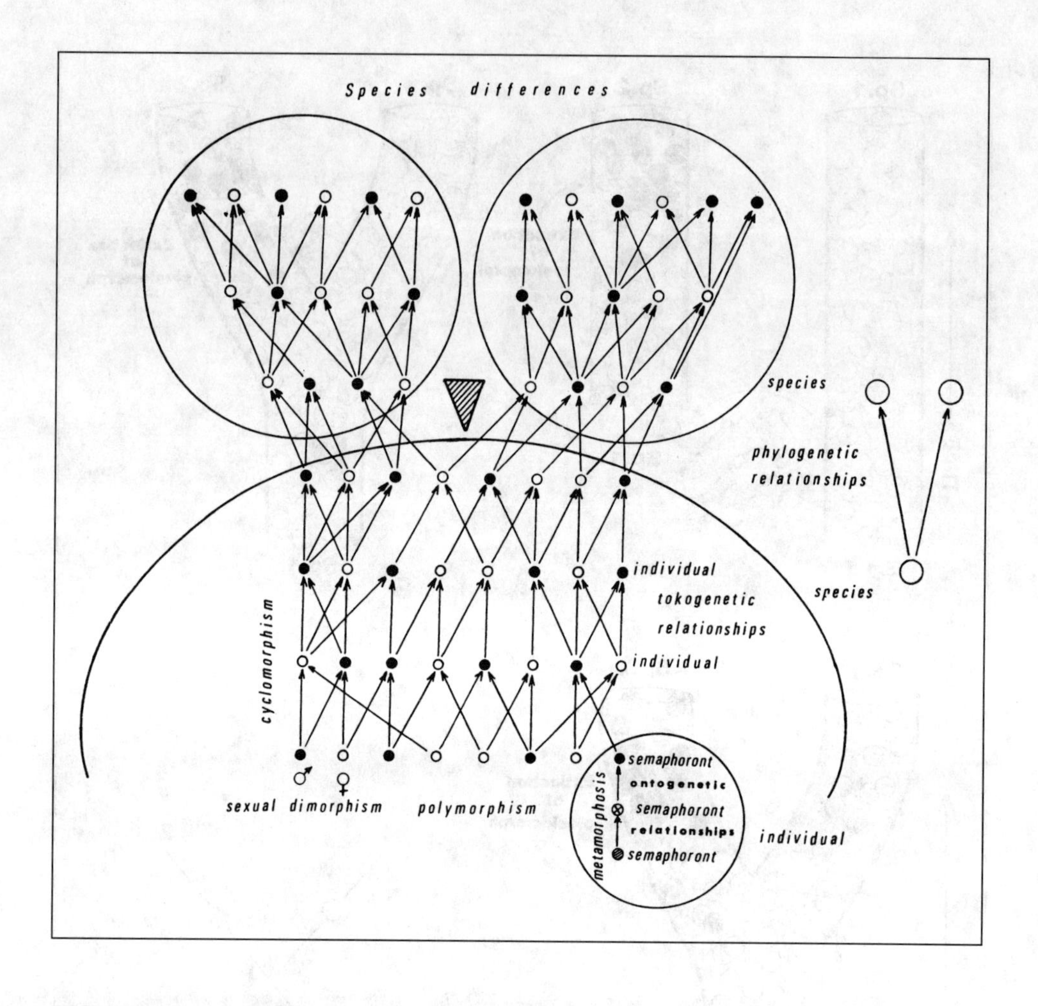

FIGURE 5.4.
Tokogenetic versus phylogenetic relationships. (From Hennig 1966, by permission of the University of Illinois Press.)

vary or that other attributes of the organisms are any less variable. The advocacy of a character-based species concept (some would say *morphological*, although *holomorphological* would be more accurate in the sense of Hennig 1966) is sometimes viewed with fear as an atavistic version of typology: "Defeated by these difficulties a few recent authors have gone back to a purely typological, morphological, species definition. This may be acceptable procedure for a museum or herbarium curator who would like to have unambiguous names on his specimens and collection cases. The authors who do so forget, however, that the species is the unit of interaction in many biological disciplines, as in behavioural biology and ecology" (Mayr 1988c:34).

The use of our Phylogenetic Species Concept is potentially more easily testable because it is based on observable characters, and it is more compatible with phylogenetic theory because speciation events are marked by character transformations. It does not, however, sidestep the kinds of rigorous comparative studies necessitated by any credible concept. Rather than losing sight of the need for biologists to have access to the units of evolution, our Phylogenetic Species Concept more precisely distinguishes them. Rather than confusing tokogenetic units with phylogenetic units, our Phylogenetic Species Concept gives to biologists a framework that avoids rampant subjectivity. And rather than confounding traits and characters within a species, the concept provides a context within which shifting gene frequencies can be explained without losing the unique historical information imparted by character transformation.

The fact that backbones are constantly distributed among all members of the terminal lineage we call Vertebrata does not imply that the backbone of the mouse and that of the giraffe are identical. But to say that because the giraffe backbone looks so different this character is not shared by the mouse is to at once deny evolutionary history and foreclose our ability to retrieve that history. We doubt that this is what Mayr had in mind. Rather, we suspect he might have had in mind cases where two species might be diagnosable, at the moment, only by differences in, for example, their courtship behaviors. If those differences can be shown by resampling to diagnose the taxa, then both species fit our Phylogenetic Species Concept (even if only one of the species' behavioral repertoires includes autapomorphic elements).

Character transformation—the fixation of a character through the removal of ancestral polymorphism by extinction events (see Nixon and Wheeler, 1992a)—provides the evidence of speciation. Because characters are modifications of preexisting characters, such transformations result in retrievable, hierarchical sets of nested attributes (Platnick 1979) (figure 5.5). By hierarchy, we refer to a situation characterized by a unique beginner element, progressing unidirectionally to many elements (Woodger 1937).

Whereas geography provides information used to recognize populations (Rosen 1978), phylogenetic species are gauged by the distribution of characters. Whether populations are allopatric, parapatric, or sympatric is far more critical to a species concept that is based on interbreeding, actual or potential. Phylogenetic species are compatible with any mode of speciation that is not in conflict with either observed character distributions or the results of cladistic analysis.

When extinction events terminate ancestral polymorphism and character transformation is complete, the pattern of sister species relationship is the same regardless of the mode of speciation. Consider figure 5.3. Species 1 and 2 are diagnosed by unique sets of characters (figure 5.3D), regardless of whether their origin involved allopatry and dispersal (figure 5.3A), vicariance and subsequent allopatry (figure 5.3B), or sympatry (figure 5.3C).

FIGURE 5.5.
Character transformations fulfill Woodger's (1937) expectations of a hierarchical set of relationships, having a unique beginner (A, 1), progressing from one to many. Because characters are modified from ancestral characters and are hierarchical, they may be viewed as a nested set (B; see also Platnick 1979). The distribution of characters on the taxon cladogram is similarly hierarchical (C).

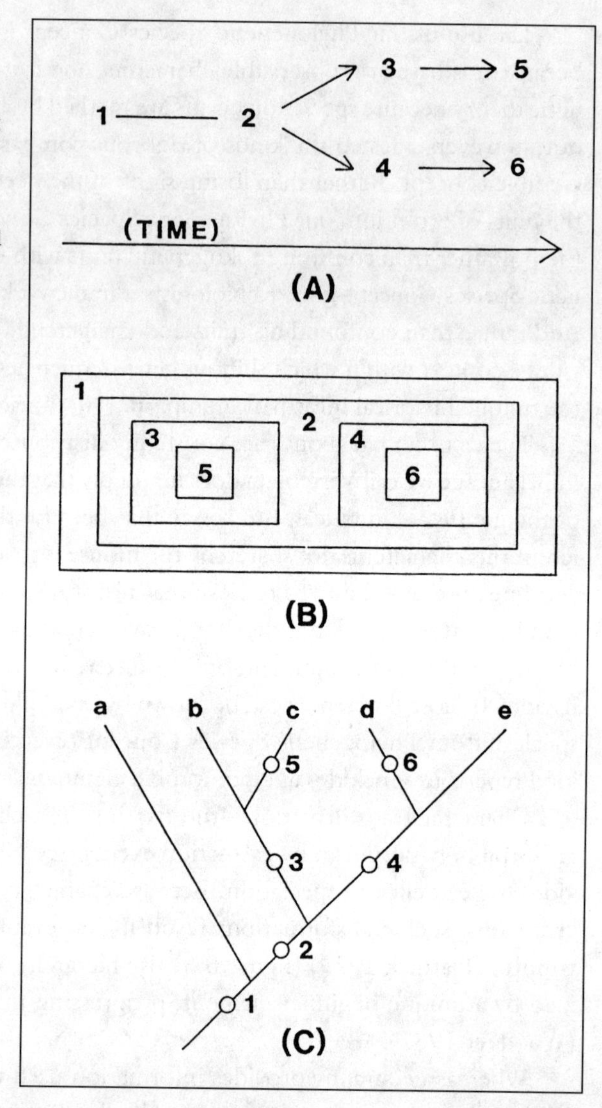

Species as Elements of Cladistic Analysis and Testable Hypotheses

What are the basic units that can be analyzed phylogenetically? How can such units be recognized prior to a cladistic analysis, so that the analysis may be undertaken? Phylogenetic species are precisely such elements. That is, phylogenetic species are the smallest groups of organisms among which historical patterns of common ancestry may potentially be retrieved (Nixon and Wheeler 1992a) and which may not be divided into smaller units with similar properties.

Based on the observation that an aggregation of populations shares a unique combination of character states, we hypothesize that it is a distinct species. With this hypothesis come two kinds of implicit predictions that may be tested through additional observations. First, the specified characters are in fact shared by all comparable semaphoronts included within the species. If additional observations reveal that supposed characters are in fact distributed as traits, then our hypothesis is falsified and we have overestimated the number of species. Second, none of the putative traits of subdivisions of the species are themselves constantly distributed characters. If further observations reveal that supposed traits, or previously unstudied attributes of one or more populations within the species, are in fact characters, then the species hypothesis is falsified and we have underestimated species diversity.

The subjectivity of polytypic biological species is avoided because the concept clearly mandates under what circumstances a population or group of populations is to be recognized as a species. This is more than a convenience of nomenclature, because these species are also the end-products of evolution, as unambiguously marked by character transformation. These characters are the evidence of an evolutionary pattern in need of a process explanation.

Unit Species and Monophyletic Groups: Describing Biological Diversity

Species are generally regarded as the fundamental units of formal scientific nomenclature. The purpose of such names is to facilitate the storage and communication of facts and knowledge about organisms and to express succinctly predictions about character distributions. The objectivity and precision afforded by phylogenetic species for recognizing the end-products of evolution impart a fine level of resolution for such communication. Where such species tease apart evolutionary products at a level too fine for the purposes of talking about particular properties of organisms, monophyletic groups give biologists the necessary language to make more sweeping statements about groups at a coarser level of resolution.

Increased awareness of an imminent anthropogenic mass extinction event (Wilson 1985, 1988, 1992) has heightened concern about maintaining biological diversity. This concern is rapidly being translated into government actions, such as the Convention on Biological Diversity, a systematics agenda (Systematics Agenda 2000, 1994a, 1994b), and the formation of a United States biological survey in the Department of the Interior (Raven et al. 1993), among others (e.g., Wheeler 1995b; Wheeler and Cracraft 1996; AMNH 1999 and references therein).

Alarmingly, many decision makers are unaware of the central role that taxonomy plays in biodiversity studies. Conservation literature makes superficial reference to species diversity and generally refers to simple inventories of the number of biological species reported from a given area. Geographic regions are considered to be more biologically diverse when a longer species list can be documented. This approach has two fundamental flaws, both of which are related to species concepts, and confuses what Nixon and Wheeler (1992b) have termed *species diversity* and *phylogenetic diversity.*

First, because biological species are used for most of these counts, there is no way to compensate for the uneven, subjective boundaries assigned to species and subspecies. Our Phylogenetic Species Concept, in contrast to the Biological Species Concept, would provide an objective basis for enumerating the number of diagnosable end-products of evolution present in each area. These are presumably the unique kinds of organisms that conservationists seek to preserve.

Second, phylogenetic diversity provides a measure of how documented species diversity is distributed among monophyletic groups. Indices based on phylogenetic diversity, such as that proposed by Nixon and Wheeler (1992b), have the potential to scientifically prioritize areas for conservation. Such phylogenetic information, of course, ultimately rests on the recognition of the elements of cladistics, phylogenetic species. Knowledge of phylogenetic diversity is critical to informed decisions about what species diversity means. Consider, for example, two areas A and B with an equal number of species. If species in area A are all closely related members of a single monophyletic group, while those in area B represent a large number of distantly related monophyletic groups, then the conservation of area B results in saving more biological diversity in terms of the kinds and (presumably) disparity of attributes of the species (Wheeler 1995a). The latter is relevant, of course, to the diverse societal needs for which biological diversity has potential value.

Our Phylogenetic Species Concept meets the diverse requirements placed upon the species unit in biology. It makes measures of the end-products of evolutionary processes and units of formal scientific nomenclature coincident to the degree that available knowledge allows. Our Phylogenetic Species Concept provides the finest level of resolution of kinds of organisms that can be justified on the basis of constantly distributed, observable attributes and theoretically justified interpretations of hierarchical patterns. It provides the elements for cladistic analyses without requiring that such analyses be performed a priori. Phylogenetic species are formulated as hypotheses that make very specific predictions. These may be critically tested through further observations. Because phylogenetic species are independent of assumptions about specific processes of evolution and modes of speciation, they provide the logical basis from which such mechanisms may be studied. Phylogenetic species are character based, making them both theoretically sound and easily applied in practice. Furthermore, as unit species, phylogenetic species provide an objective basis for measuring and comparing biological diversity.

Phylogenetic species are fully consistent with phylogenetic theory, yet they are formulated in a way that is independent of cladistic analysis. Phylogenetic species resemble traditional morphological species in biology in their ease of recognition, objective basis in observable attributes, and openness to further observation. They differ in having a firm basis in phylogenetic theory, in being founded on a distinction between characters and traits, and in being explicitly, rigorously testable.

It has been suggested that progress in taxonomy was stifled when architects of the New Synthesis succeeded in shifting the focus away from species and phylogenetic relationships among species and toward research at the population level (e.g., Wheeler 1993, 1995c). This confusion remains in evidence today in journals and patterns of federal funding for research. Phylogenetic theory has succeeded in returning a clear focus and mission to systematic biology and in drawing the necessary distinction between tokogenetic and phylogenetic studies. In a parallel way, species concepts were diverted by the rapid advances in genetics early in the twentieth century. Rather than seeking a way to distinguish species and using genetic research to study phenomena that take place within species, species were defined in terms of modes of speciation and favored processes of microevolution. Instead of independent measures of the end-products of such processes, species became self-fulfilling prophecies of such mechanistic explications. Our Phylogenetic Species Concept returns the criteria for recognizing species to taxonomy, where it traditionally and logically belongs, and produces hypotheses that ultimately serve population researchers better by giving them an independent view of the patterns they seek to explain. These and the other benefits of our Phylogenetic Species Concept can only be realized, however, when the concept is applied in a rigorous and consistent way.

6

The Evolutionary Species Concept

E. O. Wiley and Richard L. Mayden

We view the Evolutionary Species Concept as identical to the species-as-lineages concept that is a central part of Hennig's (1966) philosophical development of phylogenetic systematics. We assert that competing concepts, including the Biological Species Concept, the several versions of the Phylogenetic Species Concept, and the concept that species are the same kind of taxa as supraspecific taxa, among others, do not serve to satisfy the basic objective in studies of biodiversity (accurate estimates of the number of species present) or the general needs of the phylogenetic system as Hennig (1966) conceived that system and as it is used by phylogeneticists today. Our purpose is to briefly discuss the conceptual development of the Evolutionary Species Concept, characterize the concept, discuss its characteristics, and then provide a justification for its use in phylogenetic systematics. We will follow these sections with a general discussion comparing the strengths and weaknesses of other concepts relative to the Evolutionary Species Concept.

Conceptual Development

The Evolutionary Species Concept predates the general awareness of, much less the acceptance of, Hennig's (1966) phylogenetic systematics. Hennig (1966) claimed that the concept of species as common descent communities could be traced at least back to Buffon (1749). Although Hennig (1966) rather liked the Biological Species Concept, he actually treated species as Zimmermann (1931, 1943) did, as individualized lineages (see Donoghue and Kadereit 1992 for additional insights on Zimmermann's contributions to phylogenetic systematics). Although the term *evolutionary species* may predate G. G. Simpson, it was he who championed it. Simpson (1951, 1961) wished to formulate a species concept with a temporal dimension, an alternative to the nondimensional Biological Species Concept of Dobzhansky (1937) and Mayr (1942, 1963). So far as we can tell, no one, including Simpson, took the concept very seriously. Simpson (1961:165) abandoned the concept in taxonomy. Mayr in 1963 did not mention the concept by name, and later (1969:35) only mentioned it in passing. There seem to be two reasons for abandoning the concept. The first was the mistaken idea that evolutionary continuity precluded objective delimitation of lineages and hence evolutionary species (cf. Simpson 1953:35, 1961:135; Mayr 1969:35; see also Wiley 1978; Kimbel and Martin 1993:546 for comments). The second was the idea that the taxonomy of species names need not have much to do with lineages as they appear in phylogenies (Mayr 1982:294; see comments by Kimbel and Rak 1993).

It is ironic that two phylogeneticists choose to defend a concept championed by an "evolutionary" taxonomist. How might this have happened? Wiley (1978) reconsidered the Evolutionary Species Concept and explored its corollaries. He asserted that the concept of species as lineages was particularly well suited to phylogenetic systematics. This is just what Hennig (1966:29) asserted:

> We have defined the phylogenetic relationships we are trying to present as those segments of the stream of genealogical relationships that lie between two processes of speciation. Thus by definition phylogenetic relationships exist only between species; they arise through the process of species cleavage. The key position of the species category in the phylogenetic system corresponds to the following: the species are, in the sense of class theory, the elements of the phylogenetic system. The higher categories of this system are groupings of species according to the degree of their phylogenetic relationships.

Other key characteristics of evolutionary species that were discussed by Wiley (1978, 1980a) as corollaries are also found in Hennig (1966). These include the concept of species as individuals (Hennig 1966:81–83), the corollary that chronospecies are a biological impossibility (Hennig 1966:62–65), and the rejection of morphologically based "operational" definitions of species as typological (Hennig 1966:32). Hennig (1966) was correct. Species are lineages, ontological individuals existing through time and bounded by speciation events. At least some species may be discovered through "auxiliary" means, such as morphological, behavioral, and ecological studies. Their reality, however, is not vested in the fact that different species differ in some aspect of morphology, behavior, or ecology. Rather, their reality is vested in the reality of tokogenetic descent. That there are so many or so few species is not due to morphological changes but to the disruption of preexisting tokogenetic systems and the establishment of new systems (speciation).

Wiley's (1978) analysis of the Evolutionary Species Concept coincided with a general increase of interest in species concepts and their relationship to phylogenetic systematics. Some researchers (Wiley 1981a; Ax 1987; Brooks and Wiley 1988; Funk and Brooks 1990; Brooks and McLennan 1991; Mayden and Wiley 1992) have made evolutionary species an integral part of their approach to phylogenetic systematics. It has been used extensively in studies of speciation where biologically comparable units are necessary in order for analysis to proceed (Wiley 1981a; Mayden 1985, 1988; Wiley and Mayden 1985; Lynch 1989; Funk and Brooks 1990; Brooks and McLennan 1991; Frey 1993).

There is increasing application of theoretical ideas to the world of taxonomic practice. Frost and Wright (1988) used the concept of individuality and elements of the Evolutionary Species Concept to investigate the taxonomy of parthenogenetic lizards. Frost and Hillis (1990) addressed several species-level problems in herpetology by applying the Evolutionary Species Concept. Collins' (1991) application of the Evolutionary Species Concept sparked a debate among herpetologists both for (Collins 1992; Frost et al. 1992) and against (Van Devender et al. 1992; Montanucci 1992) species as lineages. Collins (1993) applied the concept in his revision of the Kansas herpetofauna. Considerable use of the concept has been made in ichthyology. These include applications to killifishes (Wiley and Hall 1975; Wiley 1977, 1986; Cashner et al. 1992), minnows (Burr and Cashner, 1983; Mayden and Kuhajda 1989; Markle et al. 1991; Mayden and Matson 1992), darters (Wood and Mayden 1992), and anabantids (Norris and Douglas 1992), as well as to the North American fish fauna as a whole (Mayden et al. 1992). Increasing use of the concept and closely related concepts are also apparent in paleontology (see Eldredge 1993; Kimbel and Martin 1993; Kimbel and Rak 1993; Krishtalka 1993; Rose and Brown 1993), where the contrasts between different species concepts are dramatic.

In the following sections we will discuss the Evolutionary Species Concept and some of the consequences of adopting such a concept for phylogenetic systematics and biodiversity. We begin by listing our operating principles and the specific goals we wish to attain.

Our Operating Principles

1. Accuracy is more important than precision, although both should be maximized.

2. There are real biological entities existing in nature that are worth the attention of systematists. That is, some taxa are real.

3. There are pseudoentities existing in taxonomy that are worthy of the attention of biologists only to the extent that their nature is revealed. That is, some taxa are artificial.

4. For systematists, concepts and their associated definitions, characterizations, corollaries, and implications concerning taxa are useful only to the extent that they guide us toward a more accurate estimate of the kinds, numbers, and regularities of biological entities of systematic interest.

5. Precise or "safe" definitions are to be rejected despite their operational charm if they lead us to inaccurate estimates of biodiversity.

6. Phylogenetic estimates of biological diversity are only as accurate as the concept employed in accessing that diversity.

7. Discussions of the nature of taxa cannot take place in a theoretical vacuum.

Our Goals

1. To integrate taxonomy, phylogenetic systematics, biogeography, and evolutionary biology as closely as possible in our efforts to estimate life's diversity. We submit that this was Hennig's (1966) goal. We also believe that it is the goal of many subsequent phylogenetic systematists, beginning with Brundin (1966).

2. To demonstrate that if monophyletic groups have objective reality, then so must their parts (species as lineages).

3. To characterize the Evolutionary Species Concept and provide cogent justifications for the use of the concept in phylogenetic systematic research.

4. To argue that the Evolutionary Species Concept is the only concept currently capable of recognizing all naturally occurring biological taxonomic entities.

5. To show that the logical relationship that has been demonstrated to exist between natural Linnaean higher taxa and Hennig's (1966) concept of the monophyletic group can be extended to Linnaean binominals and Hennig's (1966) concept of species as lineages. (Hennig [1966] had already accomplished this feat, so our discussion is more like a review.)

The Evolutionary Species Concept

In this section we will characterize the Evolutionary Species Concept and provide an amplification and justification of the concept. We prefer to characterize the concept rather than "define" evolutionary species because we do not wish to be caught up in discussions over words rather than concepts.

Characterization

An evolutionary species is an entity composed of organisms that maintains its identity from other such entities through time and over space and that has its own independent evolutionary fate and historical tendencies.

Amplification

1. Following Hennig (1966) and Ghiselin (1974a, 1981), evolutionary species are logical individuals with origins, existence, and ends.

2. Following Hennig (1966), evolutionary species are tokogenetic entities that function in the phylogenetic system as the analog of phylogenetic entities, clades. In the general scheme of things there is a threefold parallelism. Multicellular organisms are composed of parts (cells) linked by mitosis and manifested by ontogeny. Species are composed of parts (individual organisms) linked by reproduction and manifested by tokogeny. Clades are composed of parts (individual species) linked by speciation and manifested by phylogeny. Ontogeny, tokogeny, and phylogeny are not processes. Rather, they are the outcomes of processes.

3. We graph species as lines for purposes of portraying their relationships to other species. This is a cartographic device (see O'Hara 1993). This activity acknowledges that particular species are the result of historical processes (Frost and Hillis 1990; Kluge 1990) and that we have discovered them during the course of our research. The result, as Hennig (1966) has discussed, is that we name entities that exist between the speciation events that we graph. Thus, the Evolutionary Species Concept is the species concept of Hennig (1966) and is largely to entirely the cohesion concept of Templeton (1989; see also Endler 1989), the Cladistic Species Concept of Ridley (1989), and the concept of population lineages (O'Hara 1993; De Queiroz and Gauthier 1994). It would even be the Phylogenetic Species Concept of Cracraft (1987) if that concept were not burdened with a necessary search for the smallest evolutionary unit and did not contain a mix of epistemological and ontological characterizations. It is not, however, the Biological Species Concept of Mayr (1963), the Recognition Concept of Paterson (1985), the Phylogenetic Species Concept of Mishler and Brandon (1987), the Phylogenetic Species Concept of De Queiroz and Donoghue (1988, 1990b), or the Phylogenetic Species Concept of Nixon and Wheeler (1990). Of course, some evolutionary species are surely composed of individuals who share an autapomorphy, are reproductively isolated from other lineages, share a specific mate recognition system, or are diagnosable clusters, and so forth.

4. Sexual species may show cohesion patterns such that the tokogenetic relationships among their parts (organisms) are not well correlated with, or are uncorrelated with, any hierarchical relationships that might exist among those parts (organisms and/or demes of organisms). Contrary to the concerns of Donoghue (1985), this is to be expected (Hennig 1966; Kluge 1989).

5. Asexual species are similar to multicellular individual organisms in being composed of tokogenetic clone vectors descended from a single ancestor. Asexual species have tokogenetic relationships that are identical to their relationships of descent (Frost and Hillis 1990), just as the cells of multicellular organisms have mitotic (and sometimes meiotic) relationships that are identical to their relationships of descent. The un-

derlying processes are reproduction in both cases. Asexual species and individual multicellular organisms are rather like higher taxa in this regard given the threefold parallelism stated above. Individual organisms are also like many sexual species in that there is cohesion between the parts (cell adhesion in the individual organism, gene flow in cohesive species). We acknowledge that asexual species are not ontologically identical to sexual species (Hennig 1966:82–83), just as both asexual and sexual species are not ontologically identical to higher taxa (Wiley 1980a) and individual organisms.

An empirical manifestation of the similarities between asexual and sexual species relative to higher taxa can be observed when we consider attempts to recover the internal hierarchy among the parts of higher taxa, sexual species, and asexual species. In monophyletic groups, we expect that the hierarchical history among the parts (species as lineages) can be recovered to the extent that character evolution keeps up with cladogenesis (Wiley 1975) and to the extent that reticulate speciation does not obscure the pattern of synapomorphies (Funk 1985). There is no such expectation among individual organisms within asexual species or among populations within sexual species (Hennig 1966; Kluge 1989). If one could consistently derive a cladogram of the geographic parts of sexual species (i.e., populations), one could potentially split them into a number of species because there is prima facie evidence that they are independent lineages (following Hennig [1966] and his views on hierarchical/phylogenetic versus nonhierarchical/tokogenetic relationships).

In view of the relatively artificial separation between the classes *sexual* and *asexual species,* we differ from Frost and Hillis (1990) on a single point: we believe that there is room for asexual species within the Evolutionary Species Concept. They are quite right, however, in their original criticism, and we have adjusted our characterization of the Evolutionary Species Concept accordingly by dropping the word *population.*

6. The phrase *maintains its identity* is not a typological statement of similarity but a statement of individuality. Many species are composed of populations that only occasionally exchange genes, may be temporarily allopatric, or may appear to be allopatric within the time frame of the investigator's ability to view the problem (and this could last a thousand generations; see Lewontin 1974). If, however, subsequent sympatry occurs and the tokogenetic network is reestablished, then the parts had no separate identity. The uncertainty of allopatric or seemingly allopatric sister demes is a major cause of anxiety on the empirical level. It should not be (O'Hara 1993).

7. If lineages are independent, then they must have independent tendencies. That is, they are free to vary and evolve independent of their sister species. The divergent nature of the hierarchy we observe is proof of this phenomenon. Species are free to evolve or not, disperse or not, and so forth. If they have such tendencies, then on the empirical level we can expect to eventually discover differences among lineages that are the marks of their independence. The success of both character analysis and the application of the principle of parsimony is verification that this process occurs in nature.

8. To say that an evolutionary species has its own evolutionary fate is simply to say

that it is a real entity and not a figment of our imagination. It is easy to confuse statements of individuality with statements of similarity (cf. Templeton 1989:4). They are separate. The fate of an evolutionary species can be stated simply: it either speciates or eventually goes extinct. It may do both at the same time. In either case, it transitions from being an entity of process to being an entity of history.

9. Statements regarding unitary role, an original feature of Simpson's (1961) characterization, are criticized by some (cf. Mayr 1982) as unmeasurable and therefore typological (see also Hengeveld 1988). We view the term as an acknowledgment to individuals of their individuality. The term has no necessary epistemological component.

Phylogenetic Justifications for Evolutionary Species

Below, we list our reasons for advocating species as lineages in phylogenetic research. Note that all claims about process are based on observations of pattern.

1. If monophyletic groups are real, then so are species. If evolution is a real process (Darwin 1859), then monophyletic groups must really exist (Hennig 1966; see also Wiley 1989). Monophyletic groups are composed of an ancestral "something" and all the descendants of that something. Furthermore, at its origin, a real monophyletic group is composed only of a single something. If the monophyletic group at its origin has objective reality, then its sole and only part (the ancestral species as lineage) must also have objective reality because it is the higher taxon from the time of its origin (Hennig 1966) until it speciates (if it does so). And if the monophyletic group has objective reality through time, the ancestral lineage and all descendants of that lineage also must have objective reality. The goal of phylogenetic systematics at higher levels is to discover monophyletic groups (Hennig 1966). The goal of phylogenetic systematics at lower levels is to discover the lineages (Hennig 1966).

2. Ancestral species must be independently evolving lineages because synapomorphies diagnose monophyletic groups. The very fact that we can reconstruct much of the phylogenetic histories of groups is prima facie evidence that independently evolving lineages exist in nature. If the origin and spread of evolutionary novelties were not highly constrained by lineage history, if lineages freely exchanged genetic information, then the goal of reconstructing the histories of groups would be unattainable and the phylogenetic system would fail.

An additional proof that such lineages do exist is derived from the fact that phylogenetic analysis becomes complicated when lineage independence is not strictly maintained. Reticulate evolution produces confusing patterns of synapomorphy (cf. Funk 1985; Smith 1992). Of course, even in the absence of reticulate evolution, the elucidation of phylogenetic histories of some lineages can be difficult (perhaps impossible).

Not all lineages are leakproof. Horizontal gene flow through retroviruses or retrotransposons is theoretically possible, and interspecific hybridization has been shown to cause increases in variation in species (Grant and Grant 1989). Sometimes character evolution may not keep pace with cladogenesis (an assumption necessary to reconstruct a dichotomous pattern; see Wiley 1975). Some modes of speciation, such as modes II and III allopatric speciation, may result in unresolvable polytomies if rates of speciation are high and anagenesis in the ancestral species is low (Wiley 1981a). This would seem to be what happened once in diatoms of the genus *Stephanodiscus* (Theriot 1992) and may explain some of the results observed when one attempts to analyze the relationships among some species groups of cichlids inhabiting the rift lakes of Africa (Greenwood 1984).

Ancestral species must exist in the absence of spontaneous generation or special creation. Species that are ancestral to other species cannot have autapomorphies of their own because they share these apomorphies with their descendants (Wiley 1981a). That is the nature of apomorphies: the autapomorphies of ancestral species are passed down to become the synapomorphies of monophyletic groups (Hennig 1966; Ax 1987). Species concepts that require all species to have autapomorphies guarantee that ancestral species will never be found. If ancestral species always become extinct during a speciation event (Willmann 1986; Ax 1987), this amounts to arbitrarily precluding as many as $N - 1$ species from the possibility of discovery, where N equals the total number of descendant species. So, if there are 12,000,000 recent species, we will arbitrarily guarantee that we will be unable to ever discover 11, 999,999 ancestral species. If ancestral species can survive speciation events, then the number is less, perhaps much less. The Evolutionary Species Concept has no such weakness. Ancestral species may be hard to find, but we should not make them impossible to find because the concept of species employed precludes their discovery.

3. The Evolutionary Species Concept is the logical analog of the concept of the monophyletic group. Just as the concept of the monophyletic group is a strictly genealogical and "nonoperational" concept (Hennig 1966), so the concept of species within the phylogenetic system must be a strictly genealogical and nonoperational concept (Hennig 1966). This cannot be accomplished if operational criteria are woven into the concept. Such criteria reduce the concept of species to a typological construct. As we shall see, a concept can lead to testable consequences without being operational.

4. The Evolutionary Species Concept promotes a closer relationship between phylogenetics, taxonomy, and evolution by applying binominal nomenclature to comparable biological entities. All evolutionary species are comparable because they are the largest tokogenetic biological systems. They are not parts of larger tokogenetic systems, in the sense that demes or mitochondrial clones are parts of lineages, and they are not phylogenetic systems, as are monophyletic genera or families. A deme of *Fundulus lineolatus* (a topminnow) is not comparable with a deme of *Homo sapiens*, although both may be proper demes. The family Fundulidae is not comparable with the family

Hominidae because they are not sister taxa and because they have different times and places of origin. But *F. lineolatus* is comparable with *Homo sapiens* if both are evolutionary species. Thus, general phenomena associated with speciation can be studied even among nonsister species. This includes not only studies of speciation per se, but also vicariance biogeographic studies where the comparability of biogeographic units is built around the assumption that branching patterns (speciation) may be comparable because of a relationship between changes in earth history and their effects on different and relatively unrelated groups inhabiting the same region.

5. The Evolutionary Species Concept provides an ontological base for a logically consistent relationship between species and phylogenetic trees that is comparable to the relationship provided by the concept of the monophyletic group *sensu* Hennig (1966). This can be true only because lineage independence is a common phenomenon.

This particular aspect of the concept has been the focus of some literature. Wiley (1979a, 1979b, 1981a, 1987b) suggested that the number of phylogenetic trees for any one cladogram might be far fewer than suggested by others (cf. Cracraft 1974; Harper 1976; Platnick 1977c). These authors treat species and monophyletic groups of species as if they were ontologically identical (see also the concept of specieslike genera of Donoghue 1985). Wiley (1979a, 1979b, 1981a, 1987b) suggested that if analyses involved only monophyletic groups, then the number of possible trees for any given number of taxa would be the same as the number of possible cladograms. In other words, any cladogram that resulted from the analysis would be topologically isomorphic with one particular phylogeny. Isomorphy obtains because monophyletic higher taxa are specifically excluded from hypotheses that they are ancestors. Only the paraphyletic taxa of evolutionary taxonomy can be ancestors. In the system advocated by Simpson (1961), Mayr (1969), or Mayr and Ashlock (1991), there may be a difference between cladograms and trees at this level, but not in Hennig's (1966) system. Furthermore, Wiley (1979a, 1979b, 1981a, 1987b) suggested that only a few cladograms of specific topologies would be the result if ancestral species were included in the analysis (and the analysis was accurate). Finally, Wiley (1987b) suggested that Nelson and Platnick's (1981:143–147) specific example of the relationship between a tree and a cladogram was flawed and that their "second order tree" was logically inconsistent with the original phylogeny. Thus, their conclusions regarding the differences between cladograms and trees do not follow from their example.

This reasoning leads us to two important points. First, the evolutionary paradigm (descent with modification) and the phylogenetic paradigm (use of synapomorphies to discover monophyletic groups) are logically consistent. Second, hypotheses of specific ancestry (the only level of ancestry relevant to both paradigms) must pass specific phylogenetic tests that are necessary but not sufficient to pursue a hypothesis that a particular species is an ancestor. For example, if the diatom *Stephanodiscus niagarae* is the ancestor of the peripheral isolates *S. reimerii*, *S. yellowstonensis*, and *S. superiorensis*, then it should appear in a four-tomy with its descendants, just as shown by Theriot (1992).

6. To summarize, the Evolutionary Species Concept provides a biologically meaningful concept for comparing species in studies of phylogenetic relationships (Hennig 1966), coevolution (Brooks and McLennan 1991), historical ecology (Brooks 1985; Mayden 1987), biogeography (Brundin 1966; Hennig 1966), and speciation (Wiley 1981a; Wiley and Mayden 1985; Endler 1989; Frey 1993), as well as in paleontology (Eldredge 1993; Krishtalka 1993). No concept that captures only part of the tokogenetic stream or confuses tokogenetic and phylogenetic relationships can serve this function. Indeed, alternative concepts, although precise, may be inaccurate enough to lead investigators to compare incomparable "units" of analysis.

Discussion

Below, we discuss several issues we feel are important to understanding why we advocate the Evolutionary Species Concept as an alternative to competing concepts. Some of these issues are philosophical, some biological. In either case, we feel that they are critical considerations to accurate estimates of phylogenetic/evolutionary history and biological diversity.

Monophyly and the Species Question

A considerable body of literature has accumulated over alternate concepts and the question of whether species should be characterized in the same manner as higher taxa with the adjectives *monophyletic, paraphyletic,* and *polyphyletic* (cf. Platnick 1977b; Mishler and Donoghue 1982; Donoghue 1985; Mishler and Brandon 1987; De Queiroz and Donoghue 1988, 1990a; Nelson 1989a, 1989b; Nixon and Wheeler 1990). At least two versions of a Phylogenetic Species Concept are based on the notion that species should be monophyletic (Mishler and Brandon 1987; De Queiroz and Donoghue 1990a). Some critics (e.g., Szalay 1993) have claimed that species monophyly coupled with the need for species to become extinct at speciation events are central to (and fatal to) Hennigian systematics. Actually Hennig (1966) did not use terms such as *monophyletic* or *paraphyletic* to characterize species, nor did he state that species must become extinct at branching points.

We feel that the question of applying such terms as *monophyly* to species can be logically addressed in the following manner. First, are there ontological differences between species as taxa and supraspecific taxa? If not, then the adjectives should be applied, and perhaps applied to all individuals composed of parts. If so, then perhaps the differences encountered make terms such as *monophyly* and *paraphyly* inapplicable, inadvisable, or inappropriate at the level of species. Second, will the application of the term *monophyly* to species have adverse epistemological consequences? Specifically, will the application preclude discovery of species-level diversity? If so, then the application is not advisable, and practicing systematists faced with the possibility that ancestral species are included in the analysis will not be burdened with dealing with the terms (cf. Theriot 1992; Patton and Smith 1994). Rather, they can concentrate on the biology.

One can only conclude that those who apply these terms to species view species as having the same ontological status as genera or families (cf. Donoghue 1985). One source of this notion might stem from a reaction against the Mayr-Simpson view that species are real but that higher taxa are arbitrary (Mayr 1963:600–601). Another source for this notion, typified by Donoghue (1985) and Mishler and Donoghue (1982), can be viewed as frustration with a seeming lack of correspondence between actual biological characteristics of populations and currently available species concepts. They advocate a reliance upon a more operational concept by decoupling species as taxa from species as evolutionary units and treating species like supraspecific taxa.

Mayr (1963) and Simpson (1961) were wrong, of course. Monophyletic groups are real; the recognition of higher taxa need not have an arbitrary component (Hennig 1966). Unfortunately for evolutionary taxonomists, paraphyletic groups are as artificial as polyphyletic groups because the presence of either kind of group in a classification renders the classification logically inconsistent with what we know about the evolution of the group (Hull 1964; Wiley 1981b, 1987a; see also Kluge 1990; Wiley et al. 1991:92–99). Logical consistency was the one criterion that Simpson (1944, 1961) demanded of classifications relative to phylogenies and the one quality that his classifications did not have when they contained paraphyletic groups (Hull 1964). Fortunately, or unfortunately as the case may be, applying terms for the hypothesized reality or artificiality of higher taxa to the species level will not work very well because there are critical ontological differences between these two kinds of individuals. Like supraspecific taxa, species exist in nature irrespective of our ability to discover them, and they have limited longevity. Unlike supraspecific taxa, parts of species (individual organisms) participate in tokogeny. Unlike supraspecific taxa, the descent relationships among populations within species as evidenced by characters may change because character evolution during tokogenetic descent is not an irreversible phenomenon. Unlike species, supraspecific taxa must be discovered by finding synapomorphies (Hennig 1966); they are historical groups composed of lineages that have independent and separate tokogenetic processes occurring among their parts (Hennig 1966); they do not participate in natural processes (Wiley 1981b); and they have irreversible phylogenetic constraints that limit the direction of evolution of the species they contain (Riedl 1978; Brooks and Wiley 1988). The only irreversible behavior that occurs within species is a by-product of the historical constraints imposed by the epigenetic systems of their ancestors. (This is, of course, an important characteristic that allows hierarchical systems to continue through time.) Species are ontologically distinct from supraspecific taxa. Thus, we are not forced to use the same adjectives to describe the hypothesized reality of a species that we use to describe the reality of higher taxa.

Let us consider several possible outcomes of applying the terms *monophyletic* and *paraphyletic* to species. Many species are surely monophyletic; they are the terminal species with autapomorphies. Some species are paraphyletic (all ancestral species that gave rise to descendants through cladogenesis, an unknown number of species of hybrid origin whose ancestors were sister species), and some are even polyphyletic (an unknown number of species of hybrid origin whose ancestors were not sister species). If we followed Donoghue's (1985) or Nelson's (1989a) reasoning, we would judge all ancestral species as paraphyletic. We might then conclude that because paraphyletic supraspecific taxa are unnatural (nongroups), paraphyletic species are equally undesirable. This would exclude all ancestral species from the system even if we could find them. Why? Because paraphyletic groups are exactly the sort of groups that phylogeneticists reject. Indeed, an aversion of paraphyletic and polyphyletic groups is one of the few diagnostic characters of the phylogenetic movement (transformed cladists and untransformed phylogeneticists alike). Paraphyletic groups are the supraspecific ancestors of the Mayr-Simpson school and the basis for the concept of minimum monophyly (Simpson 1944, 1961; Mayr 1969). Perhaps we should reject "paraphyletic" species also? Perhaps ancestral species do not exist at all (Mishler and Brandon 1987). Perhaps they exist but are completely unrecoverable (Englemann and Wiley 1977). Neither alternative is palatable to us. The first alternative asserts that species are nominal or that evolution is a myth because ancestral lineages are necessary biological entities both for descent with modification to occur and for our ability to reconstruct those descent relationships. The second alternative admits that the phylogenetic system fails as a general reference system and that Hennig's goal of integrating phylogenetics and evolutionary biology will never be fully achieved. Ancestral species might be hard to find. But detailed studies have shown that they are not impossible to find (Prothero and Lazarus 1980; Echelle et al. 1983; Funk 1985; Frost and Wright 1988; Echelle and Echelle 1992). And, when they are found, there is a place in the phylogenetic system of classification that can accommodate them (Wiley 1979b, 1981a) while preserving a logical relationship between trees and classifications.

Following another line of reasoning, we may judge species to be paraphyletic and still accept them as entities. This "saves" evolution and avoids nominalism, but then we are forced into special pleading as we brush off evolutionary taxonomists who cannot understand why paraphyletic species are acceptable but paraphyletic genera and families are not (cf. De Queiroz and Donoghue 1988 versus Brothers 1985). We would have to abandon our original goal of treating species like genera. It is impossible to consistently treat species like genera because some acceptable (i.e., hypothesized to actually exist) species are allowed to be paraphyletic, but all acceptable genera are forced to be monophyletic.

Finally, one could avoid the problem by redefining the term *monophyly*. This was attempted by Mishler and Brandon (1987). They defined monophyly in such a way that species could not be ancestors. Only parts of species give rise to other species. Whole species cannot give rise to other species. This approach has two undesirable outcomes. First, we must sacrifice the tokogenetic relationships that naturally exist in a continuous genealogical lineage in favor of a convention of naming only part of the tokogenetic array. In this respect, the concept has the same unfortunate effect as Simpson's (1961) original use of the Evolutionary Species Concept. It also destroys Hennig's (1966) distinction between tokogenetic and phylogenetic relationships, biological patterns of descent that are empirically known to exist. Second, binominals become nominal, and species as taxa become entirely irrelevant to evolutionary biology. Species cannot be entities if ancestral species cannot exist. Defining monophyly in this way has some additional curious qualities. For example, one could argue that only parts of individual organisms give rise to other organisms. This may be technically possible, but it does not make much sense, as we shall see in the next section.

Like Liden (1990), we have a hard time accepting a "thing" as paraphyletic unless lots of other "things" are paraphyletic. Consider Ed Wiley. From one point of view, he is an individual. From another point of view, he is a group of cells. Why not apply these terms to Ed Wiley? He is, after all, a kind of group. Ed Wiley has three children. They do not reside in his body, nor are they all named "Ed Wiley." Obviously, following the reasoning of Donoghue (1985), Ed Wiley is paraphyletic. He might even be considered polyphyletic if one followed Nelson (1971). If Gary Nelson wished to apply the term *paraphyly* to individual organisms, he might assert that Ed Wiley does not exist (Nelson 1989a, 1989b). He is simply one of the nonhumans one encounters when observing groups of cells. This leaves Aaron Wiley in a fix because he is the product of two nonhumans, Ed and Karen. Yet Aaron exists. This made no sense to Nelson (1989b), and it makes no sense to us. De Queiroz and Donoghue (1990a) would accept the paraphyletic Ed Wiley; they would just claim that his primary spermatocytes are actually more closely related to Aaron than to Ed's own brain cells (see similar comments by Nelson 1989b). Mishler and Brandon (1987) would deny that Ed and Karen exist or would assert that they are only collections of cells. Aaron exists only until such time as he has children. Then he turns into a collection of cells just like Ed and Karen.

Contrary to the claims of Donoghue (1985), Mishler and Brandon (1987), and De Queiroz and Donoghue (1988), there is reason to believe that these terms should not be applied to species. Species as lineages are either tokogenetic arrays (panmictic sexual species), tokogenetic vectors (strictly asexual species), or some combination of the two (sexual species with limited gene flow, rotifers with alternating-like cycles, etc.). When one considers the array of sexual and asexual reproduction that occurs in nature, one is hard put to divide the world into sexual and asexual species. The relationships among the parts of species lie in the tokogenetic realm (Hennig 1966; see also comments by Platnick 1977b; Nixon and Wheeler 1990; Wheeler and Nixon 1990).

Monophyletic groups display no tokogeny, only phylogeny (Hennig 1966). Species as lineages can give rise to other species and thus can be the founders of monophyletic groups. Because of our inability to comprehend phylogenetic trees and their subparts without resorting to collective group names, we invent rules of nomenclature that occasionally result in a uninominal group name being applied to a single species (the monotypic genus, family, etc.). This could produce a paradox if we took it seriously, but we need not do so when we consider the situation as an instance of the imperfection of an otherwise convenient system of classification that was invented more than 200 years ago. Please do not misunderstand this statement. Family as category is a convenient way to keep track of a monophyletic group in a relational manner. A family as a monophyletic taxon is a discovered unit of evolutionary history and no less important than the hypothesis of descent with modification that it corroborates.

Monophyletic groups can give rise to nothing; they are units of history (Hennig 1966; Wiley 1980b). That is why the supraspecific ancestors that figured so prominently in the evolutionary theorizing of the 1940s to the 1960s will not do today (e.g., see discussions by Eldredge and Cracraft 1980; Wiley 1981a; Ax 1987; De Queiroz 1988; Eldredge 1989; De Queiroz and Donoghue 1990a).

Monophyly, as an adjective, need not be universally applied to all groups of "things" (Liden 1990) so long as there are cogent ontological differences between different kinds of entities. Just as *monophyly* is not a term applicable to groups of cells that are parts of single organisms (Wiley as paraphyletic), so it is not a term applicable to species just because species are composed of individual organisms and/or populations (Liden 1990). Species do not behave like higher taxa (Markle et al. 1991). Even asexual species, more like genera than their sexual classmates (Frost and Hillis 1990), are tokogenetic rather than phylogenetic entities (Hennig 1966; Frost and Hillis, 1990). Better to apply terms and phrases such as *valid, invalid, is a synonym of,* or *cannot be distinguished from* to denote the relative merits of binominals when applied to nature and leave the adjective *monophyletic* to denote natural higher taxa and *paraphyly* and *polyphyly* to denote unnatural higher taxa.

Following Nelson (1989a, 1989b) there is every reason to believe that species are taxa. However, there is no compelling reason to claim that the only valid taxa are the monophyletic ones, that is, those whose parts share a phylogenetically unique apomorphy. A named taxon is a group of organisms given a proper name in accordance with rules of nomenclature. Some named taxa are unnatural (Reptilia; *Lepisosteus sinensis* Bleeker, a mythical fish named on the basis of a drawing), some names denote only parts of species (observe the synonymy list of any widely ranging species), and some names apply to species that already have names (observe the average list of synonyms of well-known species). We do not feel compelled to restrict the term *taxon* only to those natural groups of organisms characterized by apomorphies because we do not think that all evidence resides in the form of apomorphies. Some evidence resides in the form of reproduction, as evidenced by gene flow, and some in geographic position. Some may come from topographic position on a cladogram, with synapomorphies constituting necessary but not sufficient evidence. Donn Rosen's work is particularly instructive in this regard. He named species not because of similarity or synapomorphy but because the topographic position of diagnosable entities that were nevertheless interbreeding along an intergrade zone demanded species recognition (see Rosen 1979:275–278). Thus, we see none of the apparent paradoxes that Nelson (1989b) outlined as causing real problems, not even the "paradox" of the ancestral species being the sum of its descendants in a formal Linnaean sense (Hennig 1966; Wiley 1979b, 1981a).

Empirical versus Operational Considerations

Believe it or not, the fact that the Evolutionary Species Concept has no prescribed discovery method embodied in the characterization is a strength and not a weakness. Consider the concept of the monophyletic group. The characterization provided by Hennig (1966) is "nonoperational." As it turns out, by making a few links between character evolution and cladogenesis, there are empirical ways to discover monophyletic groups. One finds synapomorphies. But even synapomorphies are not necessary, nor are they sufficient to characterize a monophyletic group (Hennig 1966; Wiley 1981a, 1989). Monophyletic groups do not exist because of synapomorphies. Monophyletic groups exist because tokogenetic relationships transition to phylogenetic relationships during speciation (Hennig 1966). Of course, synapomorphies are the evidence we need to discover monophyletic groups. If evolutionary species are the tokogenetic equivalent of monophyletic groups, we can hardly expect a useful species concept to embody some easy discovery method ("diagnosable," "has an autapomorphy," etc.). Such concepts, rather than being "process free," actually assume much more about process than a simple lineage concept. We reject this approach. In its place we suggest that just as finding synapomorphies allows us to discover monophyletic groups, finding evidence of lineage independence allows us to hypothesize that a particular group of organisms form part of an evolutionary species. Otherwise, the only diversity of life that will be recognized and considered valid will be those entities that are allowable given the particular constraints adopted by convention and embodied in a particular species definition (i.e., reproductive isolation, morphological distinctiveness, diagnosability, etc.).

The Evolutionary Species Concept embodies several biological characteristics that can be used to investigate species questions empirically. Many of these are summarized from different sources in Wiley (1981a:58–69). When we consider the richness of the empirical data that can be brought to the question of whether a group of specimens is worthy of being hypothesized as parts of an evolutionary species, we can see links between this concept and other species concepts. All the adult male members of all the populations of the fish named *Fundulus lineolatus* share a character (vertical body bar that is thin at the tips and thick in the middle versus of uniform width), whereas all the adult females share another apomorphy (thick horizontal lines on the body versus thin lines). These characters are not shared by the closest relatives of *F. lineolatus*, but they are by no means unique to these killifishes. Thus, we have evidence that these populations form at least one independent lineage relative to other fishes in the clade *Fundulus*. Following the usual practice, we would be able to diagnose this species, satisfying Rosen's (1979) species concept. It turns out that both these characters are autapomorphies, satisfying Donoghue (1985). We are at this time unable to fully resolve the relationships among populations of *F. lineolatus*, arriving at what Donoghue (1985) might call a metaspecies or what Cracraft (1987, 1992) would call an irreducible (basal) cluster. This means that *F. lineolatus* could be considered a phylogenetic species, and it is also corroboration of the hypothesis that we are working with an evolutionary species (see comments by Kluge 1989). Further investigation might show that *F. lineolatus* was reproductively isolated from its closest relatives, *F. nottii* and *F. escambiae*, because of postzygotic isolating mechanisms. If so, then we would call *F. lineolatus* a biological species (Mayr 1963). If we discovered that the populations of *F. lineolatus* exhibited considerable gene flow, then we would call *F. lineolatus* a cohesive species (Templeton 1989). The very fact of its placement on a phylogenetic tree makes it a cladistic species (Ridley 1989). What it seems to be, in fact, is an independently evolving lineage, an evolutionary species.

Of course, some taxonomic species, phylogenetic species, biological species, and other sorts of species may turn out to be perfectly valid evolutionary species. How could it be otherwise? However, some are not. Simpson (1961), Mayr (1982), and Szalay (1993) would allow a tokogenetic network to be broken into arbitrary segments, resulting in chronospecies. This divorces the taxonomy of species names from phylogenetic descent, a curious stance for evolutionary taxonomists. These "species" are not lineages and thus not evolutionary species. There is no correlation between them and the number of cladogenetic events that have occurred during the course of descent within a clade. There is no correspondence between them and speciation as that array of processes is currently perceived. Perhaps there are cogent biological reasons for recognizing such chronospecies. We have not heard that argument but will be happy to listen. Mayr (1963) would not name allopatric but differentiated populations because they are not "good" biological species. Thus, he might not recognize *F. lineolatus*. But if vicariant allopatric speciation is the dominant mode (Lynch 1989), these are exactly the sorts of species we should be studying. Cracraft (1987) would name every deme that he could diagnose. Ed Wiley could then be considered a species because 100% of his cells can be consistently diagnosed through DNA fingerprinting from other people, and his cells are a reproductive community (mitosis) existing through time. This unit seems too small, as are the demic units that might result from a strict adherence to Cracraft's (1987) Phylogenetic Species Concept (Frost and Hillis 1990). The question is, is *F. lineolatus* a lineage? More precisely, is the hypothesis that *F. lineolatus* is a lineage a hypothesis supported by data? If so, then it can be represented in a phylogenetic tree as a line. It can be represented in evolutionary analysis as an outcome of speciation. And it can be represented in taxonomy as an entity deserving a binominal.

There is an alternate scenario: we might find consistent phylogenetic structure within *F. lineolatus.* If so, we have every reason to think that *F. lineolatus* is actually composed of several different and independently evolving lineages, several evolutionary species. This happens all the time. For example, Louis Agassiz (1854) described five species of the *F. nottii* complex, including *F. lineolatus.* During a spat of lumping, fashionable in the 1950s among ichthyologists, this number was reduced to a single species with three subspecies, one of which was *F. nottii lineolatus.* This case was used as an example of the excesses of the past by Mayr (1976:258–259) to show the superiority of nontypological species concepts. In fact, it seems that there are at least five species in this complex, perhaps the same five that Agassiz (1854) named (Wiley and Hall 1975; Wiley 1977; Cashner et al. 1992). Does this example show the superiority of the unfettered brain of an anti-Darwinian thinker? Perhaps. But perhaps it shows that evolution has resulted in a minimum of five independently evolving lineages in this small group of topminnows and that anyone who analyzed the problem in some detail would have arrived at the same conclusion as Agassiz (1854). Perhaps Agassiz had a "good eye" for these fishes regardless of his position relative to the evolutionary paradigm. Or maybe he was just lucky. Unfortunately, the type specimens he examined are either lost or relabeled, so we shall never know. Other examples may prove less refractory. For example, Patton and Smith (1994) may be working with several more species of pocket gophers than those traditionally recognized if their phylogenetic hypothesis based on cytochrome-*b* sequence data continue to stand up to other character analyses.

Operationalism has never worked very well (Rosenberg 1985) and will not do when we search for a general concept of species. There are four reasons for making this assertion. First, characters are neither necessary nor sufficient to define taxa (Hennig 1966; Wiley 1981a, 1989). Thus, character-based species concepts (cf. Rosen 1979) are insufficient. Second, any concept that embodies a discovery method confuses "what things are" with "how things are discovered." Thus, we will be guaranteed only to discover those entities that can be discovered by the operation. This more or less guarantees an inaccurate estimate of the number of lineages in nature unless we are sure that we know enough about the natural processes of evolution so that we could be sure that our operations do, in fact, capture all the diversity at the species level. However, no operational concept we have seen has been proposed by investigators who would make this claim. The alternative claim would take the following form. "I'm not claiming that I know enough about evolution to specify the operations necessary to capture all the diversity of species, but I'm not really interested in doing so because my theory of knowledge does not require this." Such a statement presumably would be acceptable, but it is equivalent to admitting that the operations are not general. Third, operationalism, by its very nature, sacrifices accuracy on the altar of precision unless processes are so well known that accuracy and precision are synonymous. Fourth, operationalism is not the same as testability. If it were, Hennig's (1966) concept of the monophyletic group would not be testable.

Concepts of species, such as the Evolutionary Species Concept, may make one feel insecure because there is no easy operation to be performed (O'Hara 1993). Indeed, at the biological interface between tokogenetic and phylogenetic realms of relationship, species frequently seem rather slippery entities when compared with the security of the monophyletic group with its associated synapomorphies. But species as lineages are infinitely more preferable on conceptual grounds and do have, as part of their logical structure and correspondence to biology, testable assertions. The problem is that many of these assertions lie outside the realm of phylogenetic systematics per se and within the realms of biogeography, population genetics, and the analysis of geographic variation.

What Is in a Name?

Selecting a name to be applied to a concept is frequently a sociological as well as a scientific concern (Hull 1988). Phylogeneticists were not about to let evolutionary taxonomists have the term *monophyly.* If accuracy and precision were the only criteria, we might well have opted for the term *holophyly,* coined by Peter Ashlock (1971). Rather than adopting this term, we opted for the concept of monophyly *sensu* Hennig (1966). We did so because the original meaning of the term *monophyly* was a "family of one" rather than a "family from one." We did so despite many years of what we saw as abuse of a term, that is, minimum monophyly *sensu* Simpson (1944, 1961). As members of the community of phylogeneticists, we advocate the use of the term *evolutionary species* for exactly the same reason. The Evolutionary Species Concept embodies the species concept of Hennig (1966) and clearly provides a parallel to the concept monophyly as that term is applied by phylogeneticists. It is, so far as we can see, fully consistent with the arguments made by a variety of phylogeneticists relative to problems of classification (Wiley 1979b), the relationship between cladograms and trees (Wiley 1979a, 1979b, 1981a, 1987a), the relationship of logical consistency and classification in biology (Hull 1964; Wiley 1981b, 1987b), the relationship between species and speciation (Wiley and Mayden 1985; Lynch 1989), and the relationship between character evolution, speciation, and systematics (Wiley 1981a; De Queiroz and Donoghue 1990a). It deserves recognition, both because of its inherent strengths and its pedagogical priority.

Service to Other Disciplines and Vice Versa

It is very easy to engage in theoretical discussions as if they had no impact on other disciplines. However, we feel that phylogenetics should have an impact on related biological disciplines. Phylogenetics already has had an impact. The concept of the monophyletic group is a case in point. Sections on transspecific evolution in the literature from the 1940s to the 1960s (cf. Huxley 1940; Simpson 1953; Rensch 1959; Mayr 1963) show a great reliance on the concept of minimum monophyly and the evolution of grades as if the concept were a valid biological concept and the grades were real (De Queiroz 1988; De Queiroz and Donoghue 1990b). When we compare this literature with recent texts (Futuyma 1986; Ridley 1993), we see a distinct absence of such discussions. When we compare the analyses of the Raup and Sepkoski research group (Raup and Sepkoski 1986) with the analysis of Patterson and Smith (1987), we see an even greater role for the concept of the monophyletic group in discussions of the evolution of higher taxa and their dynamics through time. Evolutionary biologists are listening. The concept of species that we strive to attain should perform the same service.

Working systematists are incorporating newer theoretical ideas into their work. In fishes, for example, the widespread application of the Biological Species Concept during the 1950s to 1970s resulted in a considerable underestimation of the actual number of species of North American freshwater fishes because the ability to hybridize was taken as prima facie evidence that many species formed single polytypic species. A shift away from this tendency to underestimate is reflected in several editions of the Common Names Checklist (cf. Bailey 1970; Robins 1980, 1991). Polytypic species and their associated subspecies are being replaced and the number of recognized species is increasing as differentiated allopatric populations are given species status (compare Robins 1991 with earlier editions and with Page and Burr 1991 or Mayden et al. 1992). Progress is made not because operations become easier but because old goals come into focus and new goals appear. Theory and practice come together. Estimates of species diversity over time become more accurate, as is needed if we are to understand biodiversity and species dynamics through time (cf. Signor 1985). Evolution and systematics come together when species are directly related to cladogenesis and other kinds of speciation. They never make contact when we allow the arbitrary subdivision of a single continuous tokogenetic array to be subdivided just because an evolutionary novelty becomes fixed (i.e., traits become characters; Wheeler and Nixon 1990) or because we cannot imagine doing any better (Simpson 1961). Taxonomy and systematics come together as working taxonomists apply relevant concepts to their day-to-day work. Our job, as theorists, is to give them the best concepts we are capable of giving them, not easy concepts that lead to inaccurate estimates of the number of lineages, speciation events, and cladogenetic events that have occurred. Why? Because evolutionary biologists, geneticists, physiologists, paleontologists, geologists, conservation biologists, ethnologists, and a host of other biologists will use the products of the toil of taxonomists in their efforts to better understand natural processes. It is at least partly our fault if we allow them to toil in vain.

The relationships between species, speciation, systematics, taxonomy, phylogenetics, evolution, population genetics, and actual nomenclature can become transparent. This is what we should be striving to achieve, not the goal of precision or certainty within the secure blanket of a limited-knowledge acquisition system. In the case of those working under the Biological Species Concept, this limited outlook amounts to claiming that differentiated but allopatric populations are not interesting and that species-level nomenclature bears no particular relationship to phylogenetic trees. In the case of some cladists, it amounts to claiming that synapomorphies are the only data relevant to systematics. Better to strive for a truly evolutionary system where pattern and process can be studied and reciprocal illumination prevails. The Evolutionary Species Concept assists in achieving this goal.

PART 2

Critique Papers (Counterpoint)

7

A Critique from the Biological Species Concept Perspective: What Is a Species, and What Is Not?

Ernst Mayr

For someone who has published books and papers on the biological species for more than 50 years, and who has revised and studied in detail more than 500 species of birds and many species of other groups of organisms, the reading of some recent papers on species has been a rather troubling experience. There is only one term that fits some of these authors: *armchair taxonomists.* Because many authors have never personally analyzed any species populations or studied species in nature, they lack any feeling for what species actually are. Darwin already knew this when, in September 1845, he wrote to Joseph Hooker, "How painfully true is your remark that no one has hardly the right to examine the question of species who has not minutely described many" (Darwin 1887:253). These authors make a number of mistakes that have been pointed out again and again in the recent literature. Admittedly, the relevant literature is quite scattered, and some of it is rather inaccessible to a nontaxonomist. Yet, because the species concept is an important concept in the philosophy of science, every effort should be made to clarify it. I here attempt to present, from the perspective of a practicing systematist, a concise overview of the important aspects of the "problem of the species."

The species is the principal unit of evolution, and it is impossible to write about evolution, and indeed about almost any aspect of the philosophy of biology, without having a sound understanding of the meaning of biological species. A study of the history of the species problem helps to dispel some of the misconceptions (Mayr 1957a; Grant 1994).

This essay is largely based upon and incorporates extensive excerpts from Mayr (1996) by permission of the University of Chicago Press and Mayr (1988b) by permission of Kluwer Academic Publishers.

I was rather astonished that so few of the authors paid any attention to the two totally different meanings of the word *species* when referring either to species as taxa or to the category species. Indeed, several of the so-called species definitions did not define the category species at all but were simply operational prescriptions as to how to delimit species taxa. Anyone who does not clearly distinguish between these two uses of the word species will get involved in hopeless arguments.

I did not find anything in any of the papers that affected in any way what I had presented in my own contribution. Wiley and Mayden in their paper make the claim that the biological species concept "do[es] not serve to satisfy the basic objective in studies of biodiversity," but as far as I could see, they did not provide any evidence to support this claim. The defenders of the Biological Species Concept use it to determine whether certain populations can be inferred to be species taxa or not.

The Hennigian Species Concept

The Hennigian Species Concept suffers particularly from the myth that when a new species originates, the old species disappears. This is fact in the case of dichopatric speciation, where one species splits into two. It is, however, not true in the case of peripatric speciation, where a new species originates by budding, that is, by the establishment of a founder population. The phyletic lineage representing the parental species is not affected in any way whatsoever by this founding event. This is particularly true in the cases where such a phyletic lineage establishes numerous such founder species. Several of the authors have wrestled with this problem, but with no results, except for the claim that the phyletic lineage is now a new species. Because the "new" species is evidently the same genetically as the old species, I do not understand how it can be called new.

I am glad that Meier and Willmann point out that Hennig had essentially a Biological Species Concept. The weakness of his species concept, made particularly clear in his diagrams, was, as I mentioned, that for Hennig speciation was a splitting of lineages, even though in his own taxonomic researches he had shown the important role of peripheral isolates. Meier and Willmann cite my emphasis on gene flow in my 1963 book. They fail to mention that in 1970 and in all my subsequent publications I more or less refuted this early viewpoint. There is no doubt that I have since consistently reiterated what I had already said in 1957 (Mayr 1957b:379): "the essence of the biological species concept is discontinuity due to reproductive isolation."

As I mentioned above, what Hennig completely ignores is that a reproductive gap may separate a new founder species from the parental phyletic lineage, but no such gap appears within the parental lineage at the time of the speciation event.

Normally one calls a population a species when it has acquired isolating mechanisms, protecting its gene pool against its parental species or a sister species. In other words, such a species is the product of the process of multiplication of species. However, the paleontologist also encounters cases where a phyletic lineage changes over time to such a degree that sooner or later it is considered to be a different species. The occurrence of the origin of such phyletic species is usually ignored when nonpaleontologists speak of speciation. Phyletic evolution does not produce an additional entity—it merely modifies an existing one. Nevertheless, the changes are sometimes sufficiently pronounced so that the paleontologist gives a new species name to the modified phyletic lineage. Gingerich (1979), in particular, has called attention to the relative frequency of such cases. Such new species usually differ only in size and proportions, but not in the acquisition of any notable innovations. Such phyletic speciation must be mentioned because it is what paleontologists usually have in mind when they speak of speciation. It is for such species that Simpson (1961) proposed the evolutionary species definition. It has been impossible so far to discover any criteria by which a phyletic species can be demarcated against ancestral and descendant "species." It is for this reason that Hennig (1966) rejects the recognition of new species without speciation.

In his discussion of the origin of species, Hennig (1966) only considers the case of a phyletic lineage splitting by dichopatric speciation into two daughter species. He considers both daughter species as new species. He ignores the more frequent case where by budding from a phyletic lineage a new daughter species originates through peripatric speciation. By his definition, Hennig is forced to call the phyletic lineage after the budding point a new species, even though it has not changed at all. Hennig's species definition also results in difficulties when a phyletic lineage gradually changes into a new species, even though there has been no splitting of the lineage or any budding. Hennig is forced to ignore such phyletic speciation no matter how conclusive the indirect (morphological) evidence for the origin of a new species may be. On the whole, whenever biologists speak of species, they have in mind the product of the process of multiplication of species, not the product of phyletic evolution.

Meier and Willmann state that I "failed to provide a criterion that specifies how and when biospecies originate and cease to exist." Actually, I have clearly stated how they originate (by the establishment of incipient species and the completion of the speciation process), but if a phyletic lineage is concerned, indeed there is no clear-cut place where the species terminates. This is equally true of their own species definition, unless a splitting event is the termination of a species. But such splitting may apply only to a minor percentage of all species.

Meier and Willmann correctly point out that in my later species definitions I omitted the word *potential* from the definition of species. I did this because I felt that the statement of interbreeding and noninterbreeding contained implicitly the information that such populations either did or did not have isolating mechanisms. It is quite true that if one omits the capacity for interbreeding, reproductive isolation will be essentially the same as geographic isolation. It is quite interesting that Dobzhansky in his early definitions of isolating mechanisms did indeed include geographic isolation among them.

I would like to recognize that Meier and Willmann in their treatment of the species problem have a better awareness of the actual problems than any of the other authors of these papers.

THE EVOLUTIONARY SPECIES CONCEPT

Wiley and Mayden's definition of the evolutionary species suffers from exactly the same weaknesses as Simpson's definition (1961). First of all, every isolated population would become a species under this definition because it "maintains its identity from other such entities through time and over space"; furthermore, it is perfectly impossible to determine for any population whether it has "its own independent evolutionary fate [in the future] and historical tendencies." The major operational problem of the taxonomist who wants to determine the species status of an isolated population is not helped in any way by this definition.

To include in the species definition that species evolve or that they are "the result of historical processes" (species taxa) is superfluous because everything in the living world has exactly the same qualifications.

Regarding my statement that higher taxa are arbitrary, I have specified in a number of more recent publications how one has to interpret this remark. Of course, most higher taxa, let us say birds or beetles or penguins, are not at all arbitrary. What is arbitrary in many cases, however, is the demarcation of the taxa. The frequency with which higher taxa are split, combined, or have genera transferred to another higher taxon shows that higher taxa lack the usually clear-cut demarcation of species. Furthermore, the rank of higher taxa is often completely arbitrary, and the same higher taxon (group of genera) will be called a tribe by one author, a subfamily by a second author, and a family by a third. This is the arbitrariness to which I referred.

Wiley and Mayden state, "Mayr (1963) would not name allopatric but differentiated populations because they are not 'good' biological species." This is not true. I have consistently named them, but I called them subspecies.

Wherever monophyly is discussed, the treatment is marred by the self-serving claims of the Hennigians. The simple fact is that for about 90 years the term *monophyletic* has always referred to the status of a taxon. It was Hennig who shifted the term to an entirely new concept. Such a transfer of terms is simply not permissible according to all the traditions of science. There is no reason whatsoever not to accept a term for Hennig's new concept, but that term is not *monophyly*, but *holophyly*.

Every technical term has its area of application. It should not be used outside that area. For instance, the terms *apomorphic* and *plesiomorphic* are excellent choices for describing characters of taxa in a phyletic lineage. However, they are, in my opinion, nonsensical when applied to species. It does not clarify anything about the species status of a population when the term *plesiomorphic* is applied to its ability to interbreed with another population. Some 50 years ago the fact that species are not constant, but the product of evolution and still potentially continuing to evolve, was included by several authors in the species definition. For instance, in 1945, A. E. Emerson defined the biological species as follows: "a species is an evolved or evolving genetically distinctive, reproductively isolated, natural population" (p. 14). Indeed, nothing distinguishes a biological species from a natural kind better than its capacity to evolve. Yet, this is not a sufficient criterion. Everything else in living nature also has the capacity to evolve. Every population, every structure and organ, is the product of evolution and continues to evolve; genera and higher taxa evolve, and so do faunas and floras. Most importantly, the capacity for evolving is not the crucial biological criterion of a species; that would be the protection of its gene pool. It is for this reason that I and most adherents of the Biological Species Concept omit *evolving* from the species definition. Those authors who still emphasize the evolutionary aspect of the species have never made it clear what the real significance of species is.

The paleontologist Simpson attempted to make evolution the basis of a species concept: "An evolutionary species is a lineage (an ancestral-descendant sequence of populations) evolving separately from others and with its own unitary evolutionary role and tendencies" (1961:153). He replaced the clear-cut criterion (reproductive isolation) of the Biological Species Concept with such undefined vague phrases as *maintains its identity* (does this include geographical barriers?), *evolutionary tendencies* (what are they and how can they be determined?), and *historical fate*. What population in nature can ever be classified by its historical fate when this is entirely in the future?

As defended by Simpson and Wiley, I discussed the Evolutionary Species Concept in great detail in a previous paper (Mayr 1988a). I can do no better than to repeat the conclusions I reached at that time. The Evolutionary Species Concept encounters three major difficulties:

1. It is applicable only to monotypic species. It cannot account for polytypic species that contain geographical isolates because each of these isolates fulfills the evolutionary species definition, being a single lineage that maintains its identity. Thus, the definition provides no yardstick for the placement of isolated populations. When similar morphospecies are encountered at different exposures, the evolutionary concept provides no criterion that would permit a decision on whether or not they are conspecific. Rather, as with the Biological Species Concept, one has to infer from the amount of morphological difference, exactly as one does with geographical isolates in living fauna.

2. The qualification "own evolutionary tendency and historical fate" does not permit discrimination between good species and isolates. There are no empirical criteria by which either evolutionary tendency or historical fate could be observed in a given

fossil sample. Simpson himself (1961:154–160) realized this and admitted arbitrariness in application.

3. The hoped-for capacity of an evolutionary concept to help in the delimitation of chronospecies did not materialize. This is well documented by the inability of the Evolutionary Species Concept to arbitrate in the controversy between Gingerich and his followers, who believe in a frequent occurrence of phyletic speciation, and Stanley, Eldredge, and followers, who believe in a rather complete stasis of nearly all neospecies. Indeed, the most articulate proponents of the Evolutionary Species Concept (Simpson and Wiley) agree that it does not provide a nonarbitrary method for the delimitation of species in the time dimension. This is most curious because the main reason that the Evolutionary Species Concept was introduced was to deal with the time dimension, which is not considered in the nondimensional Biological Species Concept (Mayr 1988a: 323–324).

To summarize, this concept encounters three major difficulties: (1) it is applicable only to monotypic species, and every geographical isolate would, by implication, have to be treated as a different species; (2) there are no empirical criteria by which either evolutionary tendency or historical fate can be observed in a given fossil sample (Simpson 1961:154–160); and (3) the definition does not help in the lower or upper demarcation of chronospecies, even though the main reason that the Evolutionary Species Concept was introduced was to deal with the time dimension, which is not considered in the nondimensional Biological Species Concept. Indeed, Simpson's definition is essentially an operational recipe for the demarcation of fossil species.

The Phylogenetic Species Concepts

The proponents of the so-called Phylogenetic Species Concept quite openly return to the old typological species concept. For them a species is something that is different from something else. Theirs is purely operational advice on how to delimit species taxa. Nothing whatsoever is said about the biological meaning of the species concept. A further weakness of this concept is that it does not properly define what "a unique combination of character states" is. For instance, owing to the advances in molecular biology, we can now determine unique character combinations for certain molecules characterizing human races. Would the Phylogenetic Species Concepts necessitate making these races true species? Of course not! Then what is a unique combination of character states?

In order to apply cladistic principles even to intraspecific populations, that is, to the very lowest branching points, some cladists have recently proposed the Phylogenetic Species Concept. Indeed, there are now at least three versions of this concept in existence (Nixon and Wheeler 1990). This concept was first suggested by Rosen (1979:277), who proposed to consider the lowest population or population aggregate showing a new character (apomorphy) as a separate species: "a species is merely a population or group of populations defined by one or more apomorphous features; it is also the smallest natural aggregation of individuals with a specifiable geographic integrity that cannot be defined by any current set of analytic techniques." Rosen apparently considered the absence of reproductive isolation to be a plesiomorphic character and thought that the principles of cladistics would not permit him to use such a character in the delimitation of species. Indeed, this forced him to consider every distinguishable geographic isolate as a separate species. This conclusion would have required raising the population of various fishes in just about every tributary of every Central American river to species rank, for nearly all of them have some special color gene or peculiar characteristic. From this suggestion was derived the most widely accepted definition of the phylogenetic species as the smallest cluster of organisms that is diagnosably distinct from other such clusters. The phylogenetic species apparently lacks any biological significance and merely serves as "the smallest unit suitable for cladistic analysis" (Nixon and Wheeler 1990:212). To adopt this reductionist approach would lead to a massive increase in the number of recognized species in all groups with geographical variation and isolation.

One of the major weaknesses is that it is left to the arbitrary judgment of the taxonomist what he or she considers to be such a smallest aggregation or what is diagnosably distinct. Specialists of different higher taxa often disagree on what is diagnosably important. A single color difference may be acceptable for one group, whereas for the specialist of another group a major morphological reconstruction is considered necessary. Species of different higher taxa are no longer equivalent under this species concept. When every diagnosable population is called a different species, the species of a genus will become exceedingly heterogeneous. Some of them are closely related (allopatric), and others are sympatric and only distantly related.

The thinking of the cladists seems to be inspired by an endeavor to locate for each clade the ultimate stem species. However, a number of recent molecular analyses indicate that this is a vain hope. Closely related species have very similar genotypes; therefore, they have very similar evolutionary propensities. The same identical attributes may originate repeatedly in different species by parallelophyly. This has been discovered by recent cladistic analyses of the taxa included in the genus *Lemur* and of a species group of cichlid fishes in Lake Malawi. Also, as I have shown elsewhere (Mayr 1988a:325), a single parental species may give rise by peripatric speciation to a whole number of daughter species, each of them having as its stem species a different population of the parental species. There is no need to spell out the difficulties with the stem species concept caused by such situations.

The most profound criticism of the Phylogenetic Species Concept is that it has no biological meaning whatsoever. It is an arbitrary construct of the human mind, which explains why there are so many versions of the phylogenetic species definition in the current literature. It is of no use whatsoever to the person who studies species in the field.

In the phylogenetic species definition of Mishler and Theriot, great stress is placed on the monophyly of species. I presume this is legitimate if one deals with such peculiar organisms as lichens. However, in my long experience as an animal taxonomist, I do not know of a single case where monophyly has played any role in determining whether something was a species or not. To make monophyly the key point in a species definition would not be advisable in a definition to be used routinely in zoology.

8

A Critique from the Hennigian Species Concept Perspective

Rainer Willmann and Rudolf Meier

The search for a species concept that is applicable to all of life and acceptable to all biologists has a long history. Over the years, the species discussion has probably spawned more publications than any other conceptual issue in systematics and evolutionary biology. After a period of relative tranquility following the introduction of the Biological Species Concept, the number of conflicting species ideas is again increasing, and species definitions are again being vigorously debated. The position chapters reflect this development.

We have already briefly discussed some aspects of the competing species concepts in our first chapter in this volume and will here only expand on our main objections. We felt that it was most appropriate to treat the alternative species concepts separately.

The Phylogenetic Species Concept *sensu* Wheeler and Platnick

General Remarks

There is a movement within cladistics to ban all terms and concepts from phylogenetic systematics that refer to evolution. Supposedly, a new type of cladistic systematics will emerge that makes fewer assumptions about evolution and is more operational. The Phylogenetic Species Concept *sensu* Wheeler and Platnick constitutes a step in this direction, and its operational charm may indeed appeal to some readers. However, we will argue that such an operational species concept is either entirely subjective or, at least in part, based on evolutionary definitions of certain concepts, and thus not as operational as it may appear after reading the definition (see Hull 1968 for comments on operationalism). Consider a case that could easily be encountered in a lineage with a more or less complete fossil record (figures 8.1 and 8.3; Willmann 1981, 1985b).

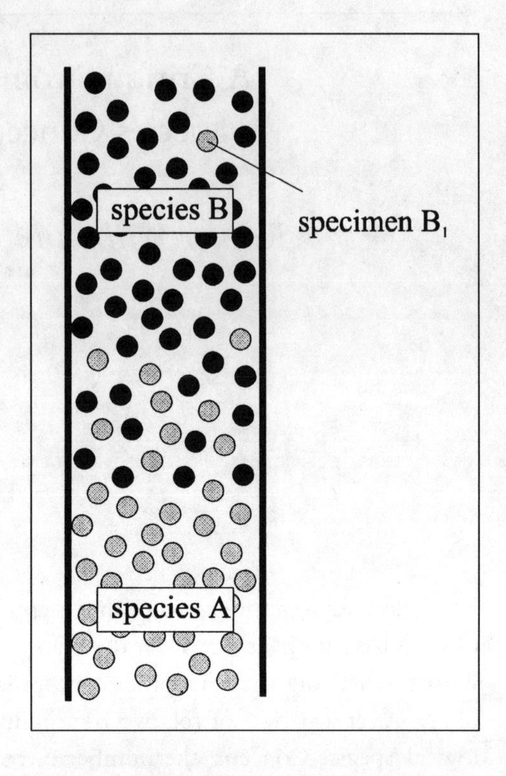

FIGURE 8.1.
Species A transforms into species B through the fixation of the shape character. B_1 has reverted to the character of species A. According to the phylogenetic species concept, B_1 belongs to either species A or species B. The first choice implies a nonevolutionary character concept and renders species A a nonnatural unit. The second choice employs an evolutionary homology concept but renders the phylogenetic species concept nonoperational.

In figure 8.1 the phylogenetic species A has a certain set of characters, and the phylogenetic species B evolves from species A through the origin and fixation of a single new character state. If this character state reverses in a member B_1 of species B, specimen B_1 can, under the Phylogenetic Species Concept *sensu* Wheeler and Platnick, be assigned to either species A or B. One could argue that B_1 belongs to species A because the two share the same combination of character states, but under this scenario species A is not a natural unit because its components originated twice. This species concept requires a nonevolutionary character concept based on similarity and a nonevolutionary definition of homology. Because there is general agreement that similarity cannot be assessed objectively, such a character concept, and, accordingly, any species concept based on such a concept, is subjective and typological.

Alternatively, one could argue that the secondary absence of the new character in B_1 is not the same character state as the primary absence in species A. Accordingly, B_1 would not belong to species A. However, this point of view is based on an evolutionary, nonoperational definition of *character* like the ones repeatedly advocated by Platnick. For example, Nelson and Platnick (1981:301) wrote, "A character is thus a theory, a theory that two attributes which appear different in some way are nonetheless the same (homologous). As such, a character is not empirically observable, and the hope of pheneticists to reduce taxonomy to mere empirical observation seems futile."

Using two character states as an example, they continued, "There seem to be only two possibilities: either one state is a modified form of the other, or both are modified forms of a third state." At least with respect to the term *modification,* Nelson and Platnick are clearly presenting an evolutionary character concept. We assume that in their species definition, Wheeler and Platnick likewise refer to an evolutionary character concept (see Davis and Nixon 1992). If so, we have to conclude that what appears at first to be an entirely operational species definition with no reference to evolution in fact requires an evolutionary character definition. Wheeler and Platnick claim that the use of the "Phylogenetic Species Concept is potentially more easily testable because it is based on observable characters." We strongly disagree. Because evolutionary characters are not observable and difficult to "test," their Phylogenetic Species Concept is not more operational and testable than competing concepts. In our example we used a reversal to illustrate our point, but the same argument applies to convergences. Because the authors do not discuss whether they adopt an evolutionary character definition, we would like them to clarify this point in their rebuttal paper.

MEANING AND DEFINITIONS OF TERMS

We are uncertain about the meaning of some terms used by Wheeler and Platnick in their species definition. They use "smallest aggregation of (sexual) populations" throughout their position chapter. If this is strictly applied, any local strain of a complex reproductive community (even a mutant strain of *Drosophila melanogaster* in a culture vial) would constitute a separate phylogenetic species if all members carry the mutation. The distinction between *trait* and *character,* which the authors consider so important, is arbitrary unless there is an objective way of delimiting their "smallest aggregation of (sexual) populations." What is a character for one aggregation becomes a trait as soon as a larger, more inclusive aggregation is considered. But which aggregation is the appropriate one? Until the authors clarify this issue, we disagree when they argue that "The use of the Phylogenetic Species Concept is potentially more easily testable because it is based on observable characters, and it is more compatible with phylogenetic theory because detectable speciation events are marked by character transformations." We have pointed out that characters are not always easily observable, at least not when an evolutionary character concept is adopted. Furthermore, as long as the status of a feature as *character* or *trait* depends on the undefined concept as the "smallest aggregations of (sexual) population" of Wheeler and Platnick, that status becomes a mere matter of opinion, and phylogenetic species are not clearly defined at all. We are therefore asking Wheeler and Platnick to clarify their concept of "aggregation of (sexual) populations" in their rebuttal paper. Unfortunately, the aggregation analysis that has been proposed by Davis and Nixon (1992) is not very helpful here. It remains unclear how the populations that are fed into the aggregation analysis are delimited. Nor are two individuals in two populations, or even a single population, identical, as is claimed by the authors (1992:430).

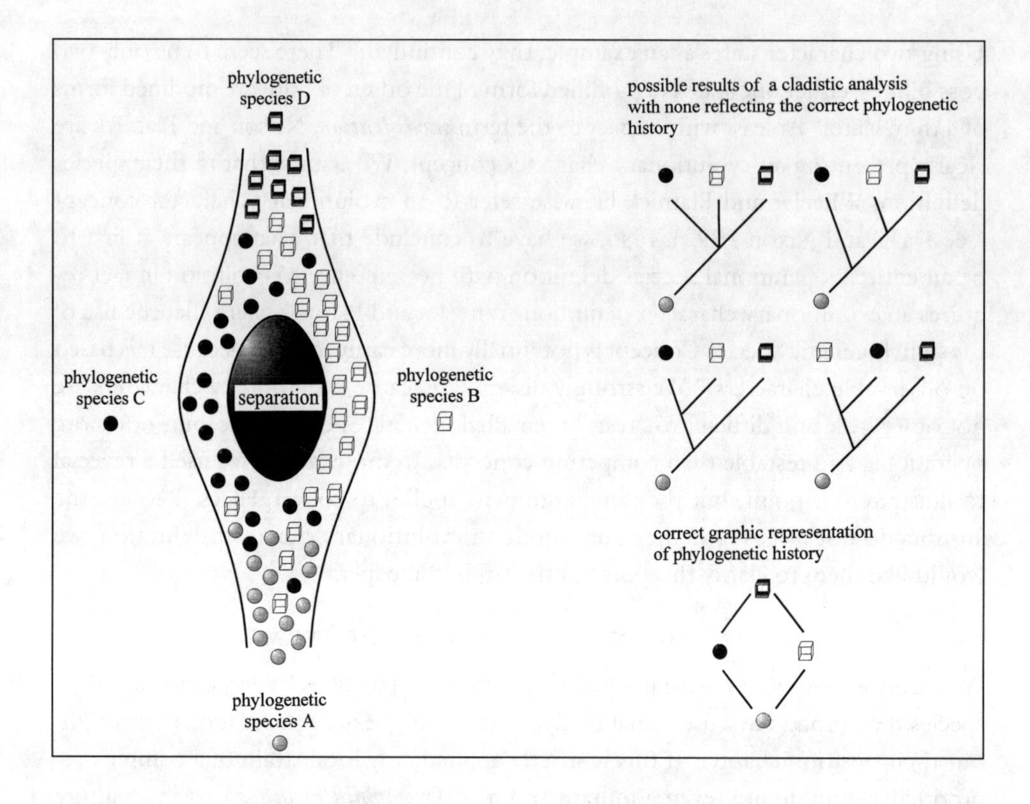

FIGURE 8.2.
The ancestral species A branches into the two phylogenetic species B and C. Species B and C were not reproductively isolated and fuse into D after the populations reestablish contact. Phylogenetic reconstruction of species A, B, C, and D cannot recover the correct history.

Of course, we would argue that only aggregations of populations that are permanently isolated reproductively should be considered species. If aggregations smaller than reproductively isolated units are chosen, they may fuse and produce a hybrid phylogenetic species (figure 8.2). If the three phylogenetic species *sensu* Wheeler and Platnick in figure 8.2 (species A, B, and C) are cladistically analyzed, their relationships are not correctly represented by any of the four possible cladograms. We must conclude that taxa like phylogenetic species A and B are not appropriate for use in cladistic analyses. Thus, we would like to know whether any gene flow between two aggregations called phylogenetic species by Wheeler and Platnick is permissible. The resulting species will be permanent tokogenetic units only if the answer to this question is "no" and different aggregations are isolated by reproductive gaps. However, if the answer is "yes," different tribes of humans (e.g., pygmies) that differ by fixed characters must be considered separate human species. According to the Phylogenetic Species Concept of Wheeler and Platnick, the Australian aborigines were definitely a separate species before the invasion of Caucasians. But they were not an end-product of evolution, as Wheeler and Platnick would argue. They have, since the invasion, begun to fuse with another alleged end-product of evolution as defined by Wheeler and Platnick's Phylogenetic Species Concept. Either we misunderstand the term *end-products of evolution,* or phylogenetic species are not such units. No species except for the ones that become extinct are truly end-products. For example, stem species that give rise to new species dissolve, but they are not end-products because the chain of generations is not interrupted.

QUANTITATIVE CHARACTERS

Wheeler and Platnick exclude quantitative characters from their discussion and effectively pretend that all characters are qualitative. This leads to a paradoxical situation because quantitative characters constitute excellent, valid characters distinguishing simultaneously living species and are used by proponents of all species concepts. The meaning of character fixation is complicated for quantitative characters, and the term *fixation* may not be applicable at all. At any one time there will be some variation in a quantitative trait, and no state will ever be fixed because there is overlapping variation with respect to this character in previous generations. Viewed along the time axis, the variation of quantitative characters is continuous. Because quantitative features never go to fixation according to Wheeler and Wheeler's character concept, they cannot delimit phylogenetic species.

FURTHER PROBLEMS WITH THE PHYLOGENETIC SPECIES CONCEPT

Potential users of Wheeler and Platnick's Phylogenetic Species Concept should be aware that, according to their species definition, the number of character fixations is equal to the number of species, because with every change a new combination of character states is produced. Even lineages that do not branch need to be divided into as many sequential phylogenetic species as there are fixations of characters (phyletic speciation). We cannot see how such sequential "species" may be considered units in a cladistic analysis because they are not different clades. On the other hand, fixations of characters are probably often not synchronous with branching events. Accordingly, some phylogenetic species are "surviving" stem species (see figure 5.2B in Wheeler and Platnick). One example can be found in Theriot (1992); another (figure 8.3) comes from Tertiary freshwater gastropods (*Melanopsis*), for which the fossil record is very complete (Willmann 1981, 1985b). As stratigraphy shows, a smooth-shelled form (a single phylogenetic species, according to Wheeler and Platnick) subsequently gave rise to several other species with differently sculptured shells. If the smooth-shelled form is considered one species surviving several branching events, it is the stem species to a number of other species and monophyletic groups. As a result, the hierarchical relationships among the taxa are obscured when a cladistic system is based on the Phylogenetic

Species Concept *sensu* Wheeler and Platnick. We have already discussed the methodological problems with surviving stem species in our first chapter and would like to refer the reader to the appropriate paragraphs. In our example, according to Wheeler and Platnick's Phylogenetic Species Concept, the monophyletic group consisting of *Melanopsis praemorsa, gorceixi,* and *inexspectata* would not be defined because their stem species is at the same time the stem species of a more inclusive monophyletic group including *M. vandeveldi* (figure 8.3).

AGAMOTAXA

Wheeler and Platnick suggest that their species concept is equally applicable to sexual and asexual organisms. We would like to suggest that the term *uniparental* be used for asexual organisms, and *biparental* (or *bisexual*) for sexual organisms, because it makes a difference whether the male sex is secondarily reduced or whether sexes are primarily missing. We do not believe that the Phylogenetic Species Concept *sensu* Wheeler and Platnick can be applied to uniparental organisms (agamotaxa), because every individual has unique characteristics and would have to be considered its own species under this concept.

Another problem with Wheeler and Platnick's concept is shown in figure 8.4. One phylogenetic species (agamospecies B) is recognized based on the apomorphic state of one character, and a second (agamospecies A) is recognized based on its plesiomorphic state. In biparental species, no problem would arise because the relationships within the species defined by plesiomorphies are not hierarchical. But in uniparental taxa the relationships of the terminals are hierarchical, and this hierarchical pattern is obscured by combining different evolutionary lineages into one paraphyletic species (see also discussion in chapter 4 by Mishler and Theriot). In figure 8.4 the agamospecies B is monophyletic, but the corresponding agamospecies A is paraphyletic.

It is a fundamental mistake to transform a hierarchical cladogram into a classification that contains one set of taxa based on the presence of the apomorphic state and an alternate set of taxa based on the plesiomorphic state. Such classifications will predictably contain a mixture of monophyletic and paraphyletic groups. Cladograms, or for that matter any hierarchically organized information, can be translated into a natural classification only by creating taxa that are nested within each other (encaptic order).

It should also be pointed out that applying the Phylogenetic Species Concept *sensu* Wheeler and Platnick to uniparental organisms will, at least theoretically, lead to as many species as there are individuals. Known mutation rates and estimates of genome size for eukaryotes allow the prediction that all individuals will differ at least with respect to a few nucleotides. Because there are no traits in uniparentals since each mutation is instantaneously fixed for this smallest of all aggregations of populations, each change will produce a new character, a new unique combination of characters, and hence a new phylogenetic species.

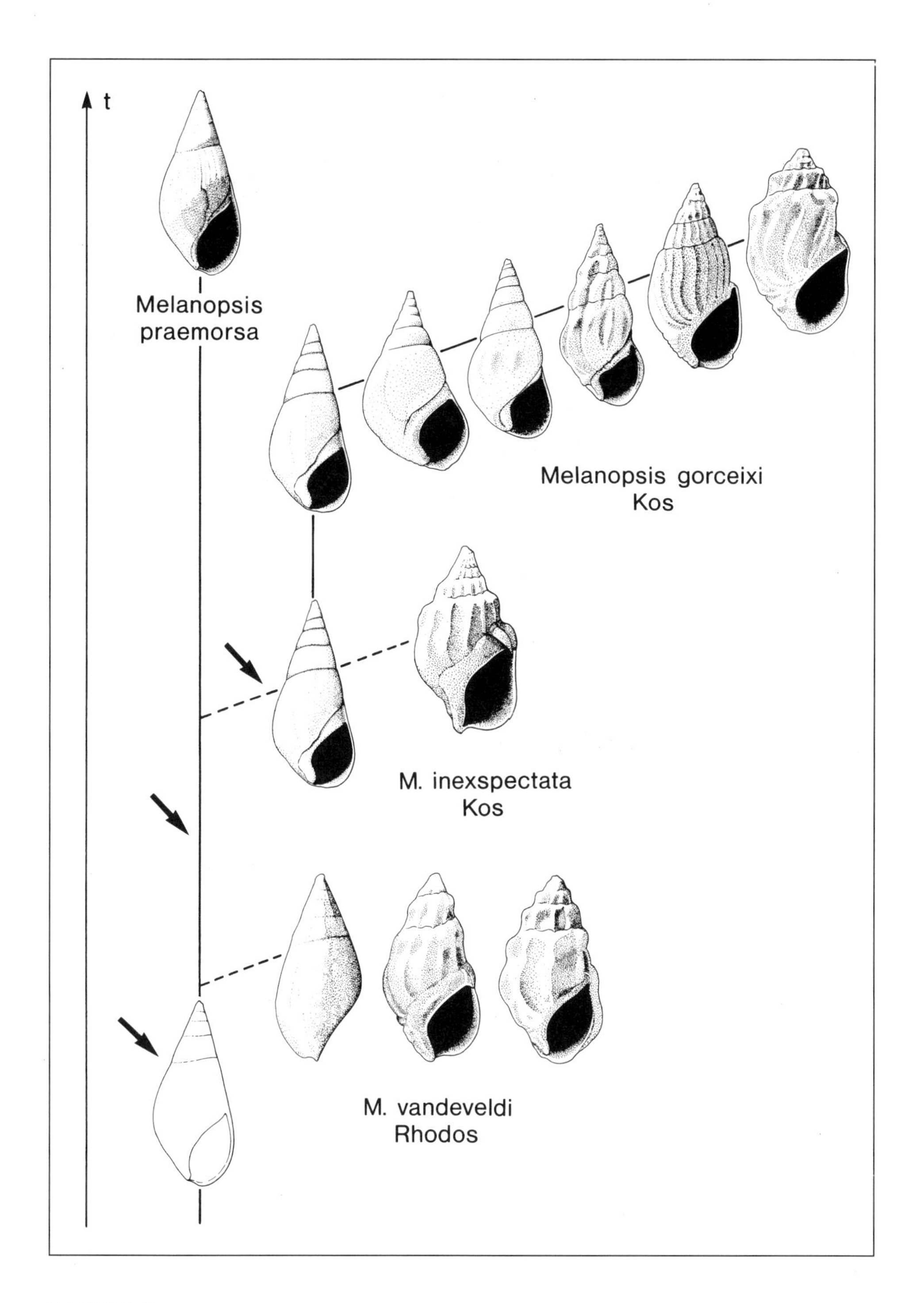

FIGURE 8.3.
Asynchronous anagenesis and cladogenesis in *Melanopsis* (see Willmann 1981). Smooth-shelled forms gave rise to three new species with independently derived sculptured shells (*M. vandeveldi, M. inexspectata,* and *M. gorceixi*). Some species concepts would consider all smooth-shelled forms as belonging to the same species, while the Hennigian concept would recognize several. Arrows point to morphologically identical stem species.

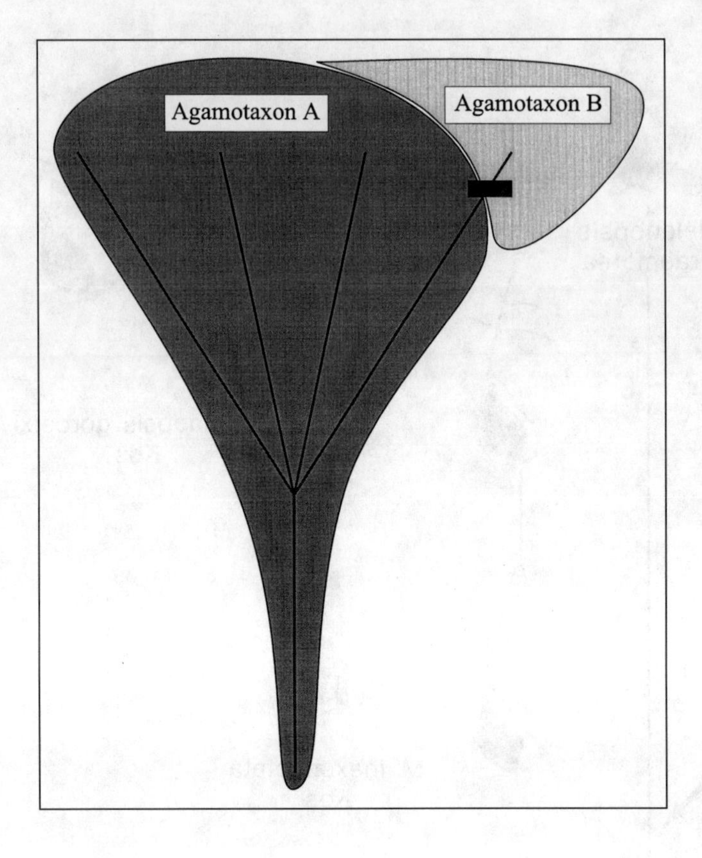

FIGURE 8.4.
The phylogenetic species concept accepts agamospecies A and B as valid species. This taxonomy obscures the hierarchical relationships within the uniparental clade because it fails to show that agamospecies A is not monophyletic.

CONCLUSIONS

Unfortunately, the virtues of Wheeler and Platnick's Phylogenetic Species Concept are difficult to assess until several terms are clearly defined. It would be necessary to know whether the authors adopt an evolutionary character definition and how they define the vague concept of "aggregation of (sexual) populations." Depending on the interpretation of these key concepts, phylogenetic species may be typological and unstable units that are inappropriate as terminals in cladistic analyses. The application of the Phylogenetic Species Concept *sensu* Wheeler and Platnick to agamospecies is also fundamentally flawed because it obscures hierarchical relationships within such clones.

The Phylogenetic Species Concept *sensu* Mishler and Theriot

General Remarks

According to Mishler and Theriot, a species concept should be compatible with phylogenetic systematics, or cladistics. Although we agree that a species concept should be rooted within the theory of evolution, we doubt that it needs to be dependent on phylogenetic systematics or, more precisely, the various phylogenetic methods as they are practiced today. If a species concept had to be dependent on a particular concept of phylogenetic systematics, each method and each variant of phylogenetic systematics would require its own species concept. It has long been realized that independent lines of evolution (clades) originate via the splitting of already existing evolutionary lineages. It follows that most organismic diversity is hierarchically structured: stem species give rise to daughter species, which in turn may become the stem species of an even younger pair of species. Hennig felt that these lineages should be termed *species,* but his conviction was independent of his phylogenetic method. If Mishler and Theriot's species concept were adopted, clades whose phylogenetic relationships were reconstructed by "reading" a more or less complete fossil record (see Willmann 1981, 1985b) could not be divided into species because species delimitations have to be based on the results of prior cladistic analysis. We believe that a species concept should be independent of the method that is used to reconstruct phylogenetic relationships. Theories about species logically precede phylogenetic analysis, and the existence of species is independent of the phylogenetic techniques used by systematists.

Redefinition of Monophyly and Phylogenetic Relationship

Much of our disagreement with Mishler and Theriot's concept results from their peculiar redefinitions of key concepts of phylogenetic systematics, including *phylogenetic relationships* and *monophyletic groups.* For example, they state, ". . . tokogenetic relationships are diachronic, ancestor-descendant connections, whereas phylogenetic relationships are synchronic, sister-group connections." Likewise their definition of monophyly is at least unusual: "*Monophyly* is defined synchronically, following the 'cut method' of Sober (1988), all and only descendants of a common ancestor existing in any one slice in time." This synchronic approach ignores stem species and extinct taxa that are, nevertheless, important to phylogenetic systematics. Stem species are as real as their descendants. Yet, they cannot belong to Mishler and Theriot's monophyletic groups. It is critically important that the definitions of key concepts such as *phylogenetic relationship* and *monophyly* are not synchronic and do not apply only to species in one time slice. As a matter of fact, Mishler and Theriot's "monophyletic groups" actually satisfy Hennig's definition of polyphyletic groups because the stem species are not included. We strongly urge that Hennig's definition of *phylogenetic relationships* and *monophyletic group* covering both extinct and extant taxa be used: "the definition of *phylogenetic relationship* is fundamental. A species *B* is more closely related to a species *C* than to any species *A*, provided *B* and *C* share at least one stem species that is not at the same time the stem species of *A*."[1]

We would also like to emphasize that, despite the sloppy use of *monophyly* in the literature (e.g., Hennig and Schlee, 1978:8), sister taxa alone do not constitute monophyletic groups. A monophyletic group always includes its stem species.

A truly phylogenetic system of all organisms should include all species that have ever existed, stem species, and their descendants. After all, every currently living species may become a stem species in the near future. For example, Hennig (1966, 1982) described the case of a species that in post-Linnaean times possibly had become a stem species. Yet, Mishler and Theriot's species concept cannot accommodate "extinct" species. Indeed, under their concept it would be justified to ignore their description as soon as they have given rise to daughter species. Phylogenetic systematics was developed for all of the organismic diversity (e.g., Hennig 1965, 1969), with stem species and the evolution of species being nuclei of the theory. With its synchronic approach, Mishler and Theriot's species concept does not have its roots within phylogenetic systematics. We are surprised they would seriously propose a concept that ignores large numbers of extinct species, including all stem species.

We are under the impression that the new definitions of *phylogenetic relationship* and *monophyletic group* are proposed to make phylogenetic systematics compatible with Mishler and Theriot's peculiar species definition. If a taxon without apomorphies cannot be considered a species, then stem "species" by definition are not species; and because they cannot be species, Mishler and Theriot decided that stem species cannot appear in definitions of *phylogenetic relationships* and *monophyly*. Thus, Mishler and Theriot redefine key concepts of phylogenetic systematics to make them compatible with their species concept instead of proposing a species concept that is compatible with phylogenetics and evolutionary biology.

TOKOGENY AND PHYLOGENY

A similarly serious problem is Mishler and Theriot's confusion of tokogeny and phylogeny. They realize that "Any cladistic analysis that fails to take into account the possibility of reticulation may not be realistic," but they propose that a cladistic analysis can be conducted using organisms or populations as terminals. Any data set can be ana-

---

[1]"Grundlegend is zunächst die Definition der phylogenetsichen Verwandtschaft (. . .). Eine Art *B* ist mit einer Art *C* dann und nur dann näher verwandt als mit jeder beliebigen anderen Art *A*, wenn sie mit der Art *C* mindestens eine Stammart gemeinsam hat, die nicht zugleich auch Stammart von *A* ist."

lyzed cladistically (i.e., ordered and forced into a hierarchical system). However, this enterprise is meaningful only if the underlying relationships are in actuality hierarchical. Within populations of biparental organisms, the (tokogenetic) relationships are netlike, and any hierarchical representation, such as a cladogram, does not actually reflect cladistic relationships. One example of such an inappropriate cladistic analysis is Patton and Smith's (1994) study, where pocket gopher "species" that hybridize were used as terminals. It comes as no surprise that under theses circumstances different genes suggest different phylogenetic relationships because the underlying relationships are not hierarchical.

Mishler and Theriot argue that "the point at which the possibility of reticulation goes to zero is well above the level at which cladistic structure can be reconstructed." As already discussed, cladistic structure below this level cannot be reconstructed with confidence when reticulation is common. Mishler and Theriot point to the work by McDade (1992), which has demonstrated that, even in the face of some noise introduced through artificially produced hybrids, cladistic analyses correctly identify certain relationships. In her study, it was easy enough to identify the hybrids and discard all relationship hypotheses that were directly or indirectly based on their inclusion in the data set, because the hybrids were known prior to the analysis. However, such identification is nearly impossible in cladistic analyses based exclusively on naturally occurring terminals. McDade pointed to positions on cladograms where hybrids are particularly likely to appear, but the same positions also may be occupied by taxa of nonhybrid origin. It is thus impossible to determine, based solely on the position of a taxon on the tree, whether it is a hybrid or not.

Mishler and Theriot's observation that "[w]ere cladistic analysis to be attempted on individuals within a panmictic group, consensus cladograms would presumably be nearly completely unresolved" is similarly useless for identifying nonhierarchically related taxa within a cladogram. Unresolved consensus cladograms also may be the result of conflicting character evidence. We doubt that the authors seriously propose that each instance in which there is lack of resolution on the consensus tree is due to tokogenetic relationships among the terminals. We do not doubt that there are hierarchical relationships below the species level, but whether a particular reconstruction only involves hierarchically related taxa cannot be known beforehand unless reproductively isolated units are used. In any case, the tools that Mishler and Theriot propose to identify taxa with nonhierarchical relationships fail to accomplish this task.

Mishler and Theriot claim that phylogenetic trees are hypotheses about nature and thus are "real" in that sense. However, this is the case only if the underlying relationships that are reconstructed are hierarchical. We claim that with organisms or populations as terminals, the underlying relationships are often netlike, and any hierarchical representation is inappropriate and does not reflect reality. The authors also claim that "Hennig's approach . . . errs . . . by postulating that there is one single breaking point at which reticulating tokogenetic relationship ends and divergent phylogenetic relationship begins." We are genuinely surprised that the authors dispute the existence of (complete) reproductive isolation. There cannot be any doubt that there is a distinct point in time at which any possibility of gene flow between two groups of organisms is interrupted once and for all. The units that are reproductively isolated may not coincide with what Mishler and Theriot conceive to be species, but strangely enough this lack of congruence leads them to question the existence of any reproductive isolation. We would like to know why they think it is not appropriate to consider reproductively isolated units as species.

Mishler and Theriot state that "breeding criteria in particular have no business being used for grouping purposes." Unfortunately, they do not specify whether they refer to interbreeding or reproductive isolation. Interbreeding is indeed of no relevance, but reproductive isolation is critical for any species concept that acknowledges the fundamental difference between reticulate and hierarchical relationships. Reproductive isolation keeps natural units permanently separate and creates natural taxa. Contrary to Mishler and Theriot, we consider this breeding criterion essential for understanding the structure of the organismic world.

MISCELLANEOUS CRITICISMS

We have problems with certain statements in Mishler and Theriot's species definition. They argue:

1. "A species is the least inclusive taxon recognized in a formal phylogenetic classification." This requirement renders Mishler and Theriot's definition arbitrary because what constitutes the least inclusive taxon depends entirely on the resolution of the cladogram, which is a function of character choice. Also, the taxon selection by the investigator will have a profound influence on which group will be recognized as the least inclusive taxon.

2. "As with all hierarchical levels of taxa in such a classification, organisms are grouped into species because of evidence of monophyly." But monophyly does not exist in the absence of hierarchical relationships, and we have shown that Mishler and Theriot will frequently be applying hierarchical representations (cladograms) to terminals whose relationships are actually netlike. Also, stem species necessarily lack apomorphies. For example, even if the stem species of the insects does not fit Mishler and Theriot's definition, it nevertheless existed, and its members must have belonged to some species.

3. Mishler and Theriot favor ranking "the smallest monophyletic groups deemed worthy of formal recognition" as species. As has been pointed out repeatedly (e.g., Ax 1987; Lauterbach 1992; Willmann 1997), ranking of taxa is nonsensical in phylogenetic systematics. The rules of nomenclature need to be changed accordingly. Furthermore, what is "deemed worthy" of formal recognition is arbitrary, and different authors will undoubtedly come to different species delimitations. Mishler and Theriot apparently

agree with this assessment when they state: " . . . they pointed out the obvious non-correspondence between groupings of organisms defined by different criteria. That is, an ecologically coherent group may be either less or more inclusive than the actively interbreeding group, and neither may correspond to a morphologically and/or genotypically coherent group." Because Mishler and Theriot do not specify which criterion should be used to rank taxa as species, each one is permissible. Apparently, the authors are even inclined to use questionable tools such as bootstrap percentages and decay indices for this purpose. Using their Phylogenetic Species Concept, the species diversity of different plant and animal taxa cannot be compared objectively because the species delimitations themselves are dependent on the subjective choice of a ranking criterion. We must conclude that biodiversity cannot be assessed using Mishler and Theriot's phylogenetic species.

METASPECIES

Absent from their discussion is the "metaspecies problem." Consider a taxon that is characterized by at least one autapomorphy and has been deemed "worthy" to be considered a species. How can one deal with an alleged basal polytomy of several terminals on the cladogram? (See figure 8.5.) Traditionally, proponents of Mishler and Theriot's species concept have argued that the "basal assemblage" could be called a metaspecies (e.g., Donoghue 1985). Our first objection is that this treatment is not in accordance with their own species definition because the assemblage lacks an apomorphy. Under their definition, the basal lineages belong to no species. We understand that proponents of such a concept nevertheless think that they have to be named somehow, but creating a new species category is hardly a solution. An obvious flaw of Mishler and Theriot's species concept is that, as soon as one taxon is ranked as a species, the rank of the corresponding sister taxon becomes unclear because it lacks an apomorphy. Under their own definition, it cannot be considered a species. Obviously, Mishler and Theriot's species concept cannot deal with this common situation.

CONCLUSIONS

Mishler and Theriot's species concept has serious flaws, even though the authors have tried to redefine key concepts of phylogenetic systematics to make them compatible with their species definition. They proposed new definitions for *monophyly,* which in fact describes a special case of polyphyly, and *phylogenetic relationship,* neither of which can deal with extinct taxa and stem species. As a consequence, a large proportion of distinct groups of organisms are unavailable for cladistic analysis because, in their view, they do not even belong to species. Also, their confusion of tokogeny and phylogeny leads to cladograms that do not reflect cladistic relationships, but merely the most parsimonious way of sorting character information hierarchically. Such cladograms will be commonly presented as phylogenies by proponents of Mishler and Theriot's Phylogenetic Species Concept.

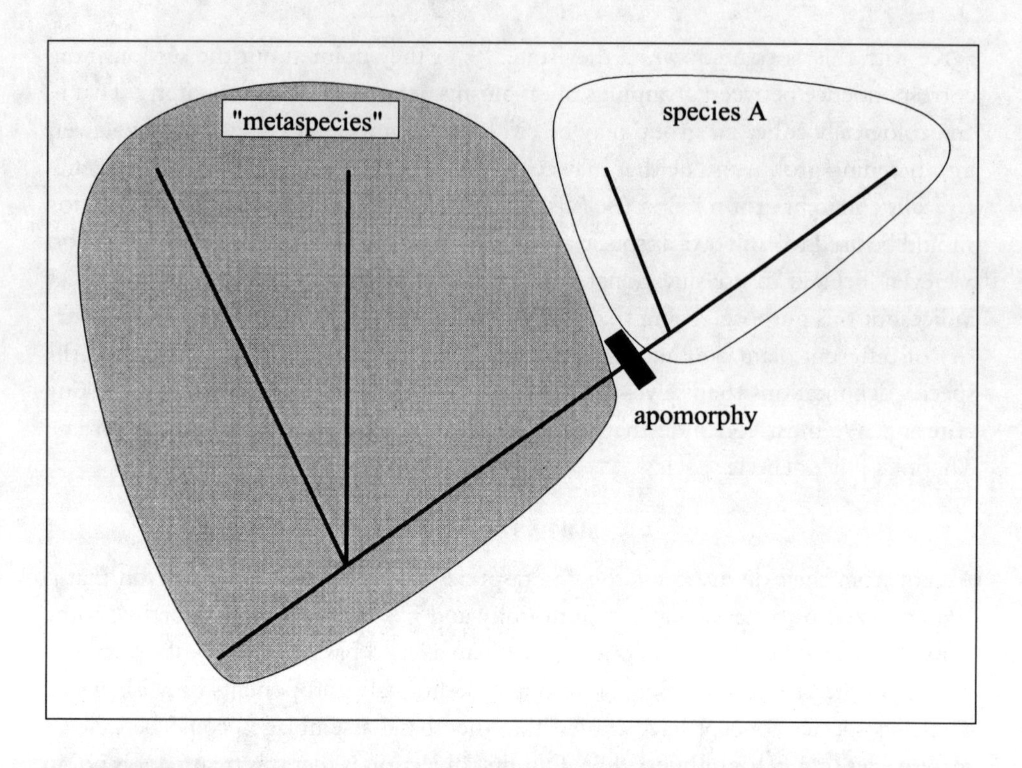

FIGURE 8.5.
The autapomorphic species concept recognizes species A. Classifying the "monophyletic" clade as a species leaves a "paraphyletic" basal rest group consisting of individuals that are not members of any species.

The Biological Species Concept *Sensu* Mayr

GENERAL REMARKS

According to Mayr, species can be defined as either "groups of interbreeding natural populations that are reproductively isolated from other such groups" or "a reproductively cohesive assemblage of populations." Apparently he has overlooked the profound differences between (1) reproductively isolated units and (2) groups of interbreeding populations and reproductively cohesive assemblages of populations. Interbreeding and cohesion are observed in many populations and demes that would nevertheless not be considered separate species under any species concept. Thus, as already pointed out in chapter 3, neither interbreeding nor cohesion can be used as a species criterion. We believe that reproductive isolation remains the sole species criterion for delimiting species at one point in time. Unfortunately, Mayr does not discuss whether he considers absolute isolation necessary for assigning species status to a population or an assemblage of populations.

SPECIES LIMITS IN TIME

Any species concept that is compatible with evolutionary biology must include a criterion for delimiting species in time. In this requirement, Mayr's Biological Species Concept fails. On the one hand, he repeatedly stresses that the Biological Species Concept "reflects the nondimensional situation"; that is, it does not address the delimitation of species in time. We consider such a species concept incomplete. On the other hand, Mayr asserts that "the Biological Species Concept presents us with the great advantage of providing a yardstick that permits us to infer which populations in space and time should be combined into one reproductively cohesive assemblage of populations and which others should be left out." But Mayr does not explicitly clarify what that yardstick for the time dimension might be: "It is the concept of reproductive isolation that provides the yardstick for delimitation of species taxa, and this can be studied directly only in the nondimensional situation. However, because species taxa have an extension in space and time, species status of noncontiguous populations must be determined by inference." There was an uninterrupted chain of generations from the origin of life through all existing species, but this has nothing to do with reproductive cohesion. Perhaps Mayr is referring to attempts to use reproductive isolation in the time dimension. We have already briefly discussed such futile attempts in chapter 3. Among other problems, such a species delimitation would entirely depend on the starting time plane. In any case, we wish to know what criterion Mayr uses to delimit species in time. We hope he will address this question in his rebuttal paper.

MISCELLANEOUS CRITICISMS

Mayr writes "that each biological species is an assemblage of well-balanced, harmonious genotypes and that an indiscriminate interbreeding of individuals, no matter how different genetically, would lead to an immediate breakdown of these harmonious genotypes." He also argues that "the species is a device for the protection of harmonious, well-integrated genotypes." If this were the case, little genetic variation within species would be expected, but population genetics research has demonstrated otherwise. Also, species certainly do not come into existence in order to protect the homogeneity of their gene pools. At least in the vast majority of cases, two populations acquire different features during a period of separation that happen to make them reproductively incompatible. These features originate incidentally, or due to differing selection regimes, but not in order to keep the two gene pools from mixing, as Mayr implies when he refers to species as protective devices.

We also disagree with Mayr's outlook on systematics. We believe that "the taxonomic rank of . . . isolated populations" should be of "major concern" to more than just the "cataloguer and curator of collections" and reject Mayr's statement "[t]o call such populations subspecies or full species is irrelevant for most biological investigations." Providing an accurate estimate of species diversity should be a goal of all biological investigations. Most biologists now agree that, in the face of a serious biodiversity crisis, a species inventory is the precondition for the conservation of biodiversity. Much of the pertinent information for this task will be contributed by the few remaining systematists who, among others, catalogue species and curate collections. Contrary to Mayr's beliefs, molecular systematics, while using most of the available funding for systematics, will provide few additional insights and not help with recognizing the millions of undescribed species. Not even with respect to the reconstruction of phylogenetic relationships can we agree with Mayr's statement that "Molecular biology, of course, has given us far more evidence on which to base our conclusions than the purely morphological evidence previously available to a taxonomist." The financial and temporal constraints of molecular work will preclude the techniques from being applied to most groups of organisms.

CONCLUSIONS

The Biological Species Concept as advocated by Mayr can be used neither to delimit species within a single time plane nor to provide a "yardstick" that allows the delimitation of species through time. Reproductive cohesion is found not only in species but also in many subspecific taxa, including populations and demes. According to Mayr's definition, these units could be called species. Also, it remains unclear how species could be delimited in time using Mayr's Biological Species Concept.

The Evolutionary Species Concept *sensu* Wiley and Mayden

SPECIES LIMITS IN TIME

As we already pointed out, to be useful for phylogenetics, a species concept must include a criterion for delineating species in time. Wiley and Mayden's concept is ambiguous about such a criterion. Their statements about stem species survival appear contradictory. On the one hand, they claim to adopt the Hennigian Species Concept when they state that "species are lineages, ontological individuals existing through time and bounded by speciation events." On the other hand, we interpret the following sentence as a rejection of Hennig's proposal that stem species do not survive speciation events: "If ancestral species always become extinct during a speciation event (Willmann 1986; Ax 1987), this amounts to arbitrarily precluding as many as $N-1$ species from the possibility of discovery, where N equals the total number of descendant species." Wiley and Mayden even go so far as to claim that "Actually Hennig (1966) [did not] . . . state that species must become extinct at branching points." We could not find any statements to that effect in Hennig (1966) and request that the authors specify the exact quote. Here we present the following two citations that clearly indicate that Hennig proposed the dissolution of the stem species at each speciation event. With regard to two sister species B and C, he stated (1966:20): "If . . . we follow backward the stream of genealogical relations among the individuals of species B, and do the same for species C, at some time in the past we would arrive at one and the same species A which we would call the common stem species of B and C." Later he wrote (1966:66), "the temporal duration of a species is determined by two processes of speciation: the one to which it owes its origin as an independent reproductive community, and the one that divides it into two or more reproductive communities"—even if "one of them hardly differed or did not differ at all from the common stem-species" (Hennig 1969:43, 1981:39–40). If Wiley and Mayden permit the survival of the stem species, there cannot be any doubt that, contrary to their claims, they do not adopt the Hennigian Species Concept.

IDENTITY, INDEPENDENT TENDENCY, AND HISTORICAL FATE

Wiley and Mayden's species definition is more or less unchanged from Wiley's earlier proposals. Some of its key terms are *identity, independent tendency,* and *historical fate. Identity* refers to individuality, *independent tendency* means that evolutionary species are free to evolve independently of their sister species, and *evolutionary fate* is supposed to imply that "it [a species] is a real entity and not a figment of our imagination." In 1981, Wiley commented on his point of view as follows: "Separate evolutionary lineages (species) must be reproductively isolated from one another to the extent that this is required for maintaining their separate identities, tendencies and historical fates" (Wiley 1981a:27). The "extent . . . required" is undefined; thus, species delimitation becomes arbitrary. Different biologists will certainly disagree about the amount of gene flow that can be tolerated between units that qualify as having separate identities. Therefore, we must conclude that Wiley and Mayden's evolutionary species are not real entities but are dependent on subjective decisions. Biodiversity estimates based on the number of evolutionary species cannot be used to compare species diversity objectively, as the authors claim in their first essay. Such estimates are possible, however, if absolute reproductive isolation is adopted as the sole species criterion. Only then will evolutionary species be "comparable because they are the largest tokogenetic biological systems." As defined by Wiley and Mayden, evolutionary species are not the largest tokogenetic units. Unfortunately, the authors do not directly comment on reproductive isolation in their paper, but there will be opportunity to address this question in the last essay of the book.

AGAMOTAXA

The Evolutionary Species Concept is supposedly also applicable to agamospecies. We have already argued that hierarchical relationships can be translated into classifications without loss of information only by creating taxa that are nested within each other. Because species are not supposed to contain other species, we cannot envision how evolutionary species can be delimited in agamotaxa without losing at least some of the hierarchical information.

CONCLUSIONS

Wiley and Mayden's Evolutionary Species Concept fails to take a stand on how to delimit species in time. We believe that any species concept useful for systematics should provide an unambiguous criterion for this task. Even when applied to one point in time, the Evolutionary Species Concept cannot be used to compare biodiversity objectively because a number of different criteria for species delimitation can be used. Different systematists will undoubtedly recognize different numbers of species, and objectivity is lost.

The concept of reproductive isolation is critical to our species concept. Some of the competing concepts presented in this volume appear to attribute little or no significance to this phenomenon. We would pose to the other authors the following questions to be addressed in their final chapters.

1. Would you agree that absolute reproductive isolation (not separation) exists between groups of biparental organisms?
2. What would you call the units separated by such reproductive isolation?
3. How does the origin of a new reproductive gap relate to the origin of new species?
4. What significance does the origin of new reproductively isolated units have for the evolution of biparental organisms?

9

A Critique from the Mishler and Theriot Phylogenetic Species Concept Perspective: Monophyly, Apomorphy, and Phylogenetic Species Concepts

Brent D. Mishler and Edward C. Theriot

Major Issues at Stake

The set of position papers in this volume clearly shows the diversity of positions that can be taken on issues surrounding species concepts, even by a group of scientists who presumably share a common evolutionary paradigm. To respond to the other papers effectively, and hopefully to clarify where the differences lie, we will first summarize the major basic issues and see how each of the authors seems to stand on these and where we differ from them (table 9.1). Then, we will go through each paper in turn and respond to it in more detail.

ontology versus epistemology

As has been pointed out by Sober (1988), a complete philosophy of science must contain both a theory of what is underlying reality (ontology) and a set of operations by which knowledge can be gained about that reality (epistemology). Whereas all the contributors would seem to agree that both aspects must be present (although some are not explicit about this), the point of difference in this regard is which of these twin aspects of a complete metaphysics is emphasized. Wiley and Mayden, Meier and Willmann, and Mayr put their emphases on the ontological side, whereas Wheeler and Platnick clearly put theirs on the epistemological side. The former are more concerned with what species should be, based on a priori theories about the speciation process. The latter are more concerned with what the character evidence shows, with little explicit concern with underlying processes (except for mode of reproduction, about which they explicitly want to consider process). In our concept, we clearly differentiate between the two aspects and strive to maintain a balance between them.

TABLE 9.1

A Comparison of Papers in This Book for Certain Key Elements of Species Concepts

	Mayr	Meier & Willmann	Wiley & Mayden	Wheeler & Platnick	Mishler & Theriot
Ontology vs. epistemology	Ontology	Ontology	Ontology	Epistemology	Balance
Special reality of species?	Yes	Yes	Yes	Yes	No
Species like higher taxa?	No	No	No	No	Yes
Species in asexuals?	No	No	Yes	Yes	Yes
Species grouping criterion	Interbreeding natural populations	Natural populations or group of populations (potentially compatible reproductively)	An entity that maintains identity through time and over space	An aggregation diagnosable by a unique combination of character states	Evidence of monophyly (apomorphic characters)
Species ranking criterion	Sufficiently isolated reproductively from other such groups	Absolutely reproductively isolated from other such groups	Largest tokogenetic biological systems	Smallest such aggregation	Smallest monophyletic groups that are formally named
Cohesion by reproduction?	Yes	Yes	Yes	No	No
Chronospeciation allowed?	No	No	No	Yes	No

SPECIAL REALITY OF SPECIES

The issue of the "reality" of species is often debated, as is the meaning of *reality* itself. For our purposes here, we will assume that *reality* (or *natural*) means a belief that the units being named exist in some way independently of the investigator who discovers them. Although it appears that all contributors agree that species are real in that sense, they differ in whether the reality of species is somehow different from that of taxonomic units at other levels in the hierarchy. Unlike the other contributors, we feel that species are exactly as real as higher taxa—no more, no less (in the tradition of Rosen 1978, 1979; Mishler and Donoghue 1982; Nelson 1989a; Vrana and Wheeler 1992).

UNIVERSALITY

Whatever one thinks of the reality of species, whether or not the same concept of species can fit all groups of organisms is a different issue. Two papers (by Mayr and by Meier and Willmann) argue that nonsexually reproducing organisms do not form species by definition. Two other papers (by Wiley and Mayden and by Wheeler and Platnick) argue that asexuals do form species, although neither paper is specific about the adjustments that need to be made to their favored concepts for asexual taxa to be comparable with those formed in sexual organisms (furthermore, we argue that adjustments to their concepts do not work in either case). In our position paper (essay 4), we explicitly state in what ways species taxa will be comparable and what ways they will not.

GROUPING VERSUS RANKING

As detailed in our position paper, all concepts of species have both a grouping and a ranking criterion (either explicitly or implicitly). The grouping criterion indicates how organisms are to be put together into a group, whereas the ranking criterion indicates how the particular group thus formed is to be recognized as a species rather than some other lower or higher rank. The criteria used in a particular concept might be similar, but they are still being used for two different logical processes. We have summarized our interpretations of these criteria for each concept in table 9.1. Proponents of the version of the Phylogenetic Species Concept that we advocate (see discussion in our position paper) have been taken to task for concluding that ranking at the species level is as subjective (no more, no less) as it is at higher taxonomic levels. The important point, however, is that all concepts avoid naming formal species taxa where they might be ephemeral or temporary (e.g., small, geographically isolated populations), even when they otherwise fit the criteria of the concept. Some judgment of significance is involved. Advocates of the Phylogenetic Species Concept *sensu* Mishler and Theriot did not invent this situation; they merely called attention to it.

THE DEFINITION OF MONOPHYLY

This issue is important because of our insistence that apomorphy is essential to the discovery of species as taxa in the phylogenetic system, but there is a need for further discussion because we and others have conflated the issues of monophyly and apomorphy (autapomorphy). Mayr is not specific about monophyly in his species concept paper, apparently because he feels that it is irrelevant to the issue at hand. Three of the papers (by Wiley and Mayden, Wheeler and Platnick, and Meier and Willmann) hold fairly strictly to a Hennigian view of monophyly (which we will call diachronic), usually phrased something like "an ancestral species and all its descendant species." Those accepting this definition rightly point out that such monophyly cannot be applied to a species. Proponents of our Phylogenetic Species Concept, including ourselves, hold a second view of monophyly (which we will call synchronic) that can be phrased something like "all descendants of a common ancestor." Here a common ancestor may be a single organism, a breeding pair, a population, or another individual (Mishler and Brandon 1987). It has an epistemology similar to Sober's (1988) "cut" method. We call it the *grab* method. Pull a branch on a phylogenetic tree, and all subbranches that come off cleanly are part of the same exclusive monophyletic group. In this definition, a single branch tip is a monophyletic group. The difference between these definitions is subtle but important. However, we now see the definition of monophyly as a secondary issue to the question of what is the nature of the smallest natural lineage that can be identified under the phylogenetic system. It is the latter question we will focus on below. The short answer is that the phylogenetic system can only identify units possessing apomorphy.

PRIMACY OF SEXUAL REPRODUCTIVE PROCESSES

Two of the papers (by Meier and Willmann and by Mayr) clearly view reproductive cohesion or isolation as the major evolutionary force holding species together. Wheeler and Platnick are not explicit about any evolutionary forces, but it seems fair to say that they give special emphasis to sexual reproduction as a pattern because of their focus on avoiding reticulate tokogeny within species. We argue that sexual reproduction may cause cohesion of a species, but not necessarily. We further argue that the methodology of the phylogenetic system does not, at any level, assume reticulation to account for character similarity (but rather descent with modification) and so does not supply us with a direct methodology for discovering reticulation (or lack of reticulation) through characters. The phylogenetic system can find the least inclusive, phylogenetically natural lineages for which there is evidence (i.e., apomorphies), although it is never guaranteed to find all phylogenetically natural lineages.

CHRONOSPECIATION?

It would seem that one of the central essences of cladistics is that splitting of lineages has to be involved in speciation. Indeed, it was in part a reaction to the chronospecies of Simpson that led early cladists to reject his evolutionary systematics. Ironically, Wheeler and Platnick, following Nixon and Wheeler (1992a), have gone back to chronospeciation as a concept. We feel that such a concept confuses the very goal of phylogenetic systematics, which is to understand patterns of recency of common ancestry among clades.

Positions and Responses

In the following sections, we amplify our criticisms of and agreements with individual position papers.

THE BIOLOGICAL SPECIES CONCEPT

Mayr effectively summarizes his long-held concept of a species as a reproductive community, defined in a nondimensional context in one place at one time. The published objections to this concept are many (e.g., see Sokal and Crovello 1970; Mishler and Donoghue 1982; Cracraft 1983), so we will only summarize our (not necessarily unique) objections here.

The Biological Species Concept clearly focuses on the ontology of species, as illustrated by Mayr's conclusion that individuals are similar because they belong to the same species rather than that individuals belong to the same species because they are similar. Mayr has repeatedly observed that, other than this, there is no necessary correspondence between reproductive isolation and phenetic/genetic difference. Nevertheless, he and others often imply a correspondence between breeding groups and all other biologically interesting types of groups (i.e., those defined by monophyly or by ecological processes), a correspondence that is usually not actually verified empirically (e.g., his study of the plants of Concord [Mayr 1992a], for which he did no breeding tests, instead relying on phenetic discontinuities) and that often does not hold when verification is sought (Rosen 1978, 1979; Bremer and Wanntorp 1979; Mishler and Donoghue 1982; Cracraft 1983).

The Biological Species Concept lacks a coherent epistemology. It is untestable for allopatric populations. That is why inferences are usually made based on whether allopatric populations have reached the degree of morphological distinctness ordinarily characterized by full species even though overall similarity may only be weakly correlated with the ability to interbreed. Even for sympatrically occurring individuals, interbreeding must remain an assumption, ironically and incorrectly based on similarity in most cases. This is, as Mayr points out, a criticism of the practitioner, not the concept, but it illustrates that the sole epistemological criterion is the ability to interbreed. In the interest of furthering a desire we share with Mayr, that systematists acquire a deeper understanding of the species concepts they practice, we point out that similarity is first a test of ancestry, not of the ability to interbreed. To paraphrase Mayr, individuals are similar because they are descended from the same ancestor; they are not similar so that they may interbreed.

The Biological Species Concept is unapologetically nonuniversal. Asexual species have no meaning here, or else every individual would be a species (because asexual organisms by definition are reproductively isolated from all other organisms). Many other arguments about asexual species are amplified throughout other position papers, especially that of Meier and Willmann, so we will address these in more detail in the following sections.

The Biological Species Concept is also unapologetically nondimensional. Mayr states that he does not care whether geographically isolated populations of song sparrows are the same species or not. We must disagree. We believe, as does any systematist who is interested in describing biodiversity on the earth (the whole earth), that patterns of shared descent across space and time matter. Populations must be linked together in phylogenetic analysis and classification.

In short, given the spectrum of tokogenetic possibilities, specieshood under the Biological Species Concept describes relationships at only one extreme of the spectrum.

THE HENNIGIAN SPECIES CONCEPT

Our response to Meier and Willmann's position paper centers around their (we believe false) distinction between sexual and asexual species. Other species concepts and resulting suggested practices also make this false distinction, so these comments can be taken to apply to those as well. Meier and Willmann take the overly simplistic view that clades cannot exchange any genetic information. Although we agree there is little evidence that highly divergent clades often participate in wholesale genetic exchange, instances of reticulation through numerous mechanisms (including sex and processes akin to sex) are routinely postulated to account for incongruent character distributions, especially among protists, algae, bacteria, and other microscopic life. Moreover, sexual reproduction does not always result in complete genetic exchange, but only exchange of parts of genomes or of organelles. A wide variety of mechanisms are known or are hypothesized to explain reticulate relationships. From the standpoint of phylogenetic systematic theory, synapomorphy is the parsimonious explanation of similarity, and homoplasy requires the ad hoc explanations of reticulation, convergence, and so forth. Thus, we see no biological reason or argument from phylogenetic systematic theory to make a special exception for reticulation due to (possible) sexual reproduction.

These authors follow Hennig rather closely and argue that species are end products of evolution, that is, entities produced by nature. They are produced when a stem species splits into two (the stem species not surviving speciation). The main similarity between their Hennigian Species Concept and the Biological Species Concept is that both include reproductive potential as the main criterion (although the Meier and Willmann restatement of Hennig emphasizes isolation). The main difference between their Hennigian Species Concept and the Biological Species Concept is that the former explicitly states temporal limits of species, whereas the latter (at least as explicated by Mayr) remains inexplicit on temporal and spatial bounds of species.

Meier and Willmann's view of a separate ontology for uni- and biparental organisms flows from the common oversimplification of splitting life into sexually reproducing versus asexually reproducing varieties and the subsequent misinterpretation of what those processes mean to origin and maintenance of characters. Although diatoms are technically gamospecies, sexuality is extremely rare, probably far too rare to account for maintained similarity in *Stephanodiscus yellowstonensis* Theriot and Stoermer. Traditional population genetic models do not require reticulation, only a pool of variation to select on. This variability can be maintained without recombination; likewise, genetic, morphological, or ecological similarity or cohesion can be maintained without reticulation. If acceptable as an example of an agamospecies, Theriot (1992) is a perfect example of what Meier and Willmann say does not exist (as are several asexual species of the moss *Tortula* studied by Mishler 1990).

Meier and Willmann argue that agamospecies will be plagued by intermediates, both temporal and spatial. The temporal argument is difficult to understand with reference to their preferred concept. These phenotypes are discrete in today's time plane, but the gaps between them disappear as we proceed backward along the time axis. "Any delimitation . . . would thus be arbitrary, and it is only a historical artifact that the intermediates are not known." Not meaning to imply any correlation between phenotypic distinction and sexual reproduction, one could simply substitute the concept of interbreeding for phenotypic similarity and come to the same conclusion. Conversely, as for their concept that the stem species dissolves and the new species begin at the point in time where isolation occurs, such new agamospecies arise, and the stem species dissolves at the point of cladogenesis (in this case evidenced by evolution of a novel character state; the sister lineage, however, may or may not be recognizable—an empirical, not a theoretical, problem).

Where intermediates exist in a time plane is also not a theoretical problem. If all features are truly continuous, there are no characters, and empirically there are not multiple species, but simply one species. Attempts to force parts of a morphocline into multiple species is not a fault of any species concept (even phenetic) or of agamospecies. It is simply bad taxonomic practice; the practice has in fact been applied to sexual species as well. Cases where there are some morphological continuities (due to developmental stochasticity, environmental effects, allometry, underlying genetic causes, etc.), but truly separate lineages, are empirical problems that can be solved by analysis of molecular data and/or by refined morphological work (laboratory experiments, common garden studies, etc.). There will be some agamolineages not marked by any evolutionary change and thus not traceable by any method.

In this vein, we agree with Meier and Willmann that the epistemology of the Phylogenetic Species Concept *sensu* Wheeler and Platnick is incorrectly applied to agamospecies; that is, simply having fixed characters does not define anything if characters are not apomorphic. Wheeler and Platnick's Phylogenetic Species Concept, as we discuss below, is ontologically based on the same concept as is the Hennigian Species Concept: that phylogenetic systematic analysis and theory are appropriate only for units with nonreticulating (hierarchical) relationships. A purely agamospecies diagnosed by lack of apomorphy is a "species" of convenience only.

A final criticism of the Hennigian Species Concept is based on the practical difficulty of observing complete reproductive isolation and the lack of testability of species hypotheses. Geographically isolated populations or assemblages (or even sympatric assemblages of microorganisms) can be observed to interbreed only under artificial conditions. Because natural cues are important in many isolating mechanisms, many species that do not interbreed in nature produce fully viable, fertile offspring under artificial conditions. The Biological Species Concept, of course, has been criticized for this, but occasional breakdowns in isolation in nature pose problems for all species concepts. An important conceptual issue, which we feel greatly weakens the scientific basis of all such concepts based on reproductive isolation, is that the empirical test evades falsification through built-in definitions: lack of interbreeding or observed interbreeding in artificial conditions ignores a plethora of isolation mechanisms in nature. This is why Mayr and proponents of the Biological Species Concept argue that it is essentially a concept valid only for sympatric populations; why the Biological Species Concept and the Hennigian Species Concept must resort to subjective subspecies concepts for allopatric populations; and why virtually all diagnoses of species under these concepts are in reality character based and not based on observation of reproductive isolation/potential/actuality.

All species concepts, whether character based or reproduction based, suffer in practice from the fact that nature is messy and breakdowns in isolation (i.e., any mechanism resulting in reticulation, such as symbiosis or partial genetic exchange) occur. Only our Phylogenetic Species Concept, however, conceptually allies all forms of reticulation as ad hoc explanations for incongruent character distributions. It does not deny that reticulation exists, only that it be observed (ideally) or at least inferred from noncharacter evidence before being invoked.

THE EVOLUTIONARY SPECIES CONCEPT

The Evolutionary Species Concept, as championed by Wiley and Mayden, identifies a species as an entity composed of organisms that maintains its identity over time and space with its own independent history and fate. It is also claimed that it follows Hennig's (1966) precepts. To our reading, the latter claim is not distinctive from any other species concepts discussed in this book. The unique features of the Evolutionary Species Concept stem from a discussion of the words of the definition and corollaries drawn from that discussion. In particular, the Evolutionary Species Concept stands alone in advocating that it lacks a well-defined epistemology but that it can incorporate both asexually and sexually reproducing organisms. It also is characterized by its insistence that ancestors survive speciation events. We find, however, some clash between these claims and the ontology of the concept, and where practical discovery processes are advocated or possible, they are directly derived from Hennig's phylogenetic systematics and so are not unique to the Evolutionary Species Concept.

Wiley and Mayden argue that their version of the Evolutionary Species Concept is that which best fits the general needs and practice of the phylogenetic system of Hennig (1966). In their view, a species is "an entity composed of organisms that maintains its identify as distinct from other such entities" with an "independent evolutionary fate and historical tendencies." In itself, this concept seems to have no stronger or weaker connection to Hennig than any other species concept. The conclusions the authors draw from this concept, however, seem to oppose directly the conclusions of Hennig. Meier and Willmann in their position paper offer another reading of Hennig: that species lineages end by definition at speciation events. We believe Hennig's position on this was quite clear and is clearly accepted by Wiley and Mayden themselves when they say that "we name entities that exist between the speciation events that we graph." Yet, Wiley and Mayden claim that ancestral species can survive speciation events. Thus, the Evolutionary Species Concept is not at all the same as Hennig's concept (either the original or that championed by Meier and Willmann).

Nevertheless, the question of survival of ancestral species is important ontologically and epistemologically regardless of whether or not the Evolutionary Species Concept really represents Hennig's position. We are less concerned with agreement with Hennig and more concerned about internal consistency of arguments, so we focus on two major points: the nature of ancestral species and the nature of asexual species. We focus on these because we believe there to be severe difficulties with the ontology of the Evolutionary Species Concept as amplified by Wiley and Mayden. We also believe that there is less agreement with the overall phylogenetic system of Hennig (1966) than they claim.

Furthermore, Wiley and Mayden are quite vague about their key arguments. They make several distinctions repeatedly (e.g., accuracy/precision, sexual/asexual, tokogeny/phylogeny, species/higher taxa) without ever clarifying or justifying them. For example, they make the last distinction by claiming that "there are critical ontological differences" between species and higher taxa. This statement, on which what they say subsequently depends, is baldly stated but not defended beyond a couple of questionably relevant literature citations.

Wiley and Mayden (along with Meier and Willmann, and Wheeler and Platnick) have also confused what Hennig meant by *tokogeny.* If you look closely at what he says, and the famous diagram, it is clear that tokogeny is not synonymous with reticulation or cross-breeding. Instead tokogeny is the diachronic relationship through time between a parent and an offspring, or an ancestor and a descendant, whereas *phylogeny* is the synchronic relationship at one instant between three or more individuals (the sister-group relationship). All these authors are confusing things when they imply the synonymy. Asexual organisms have tokogeny too. These are the correct distinctions: tokogeny/phylogeny is a matter of whether one is looking at relationships through time or at one time, whereas reticulation/divergence is a matter of whether the fundamental cladistic assumption obtains.

Wiley and Mayden claim that there is no simple or direct epistemological approach to discovering an evolutionary species; we believe this is due to the ontological inexactness of their characterization of an evolutionary species and is best illustrated by a discussion of survival of ancestral species. Their characterization of an evolutionary species includes the phrase "maintains its identity . . . through time. . . ." It is only in this context that ancestral species can survive speciation. They argue that to discount ancestral species is to discount the tokogenetic continuity in a species through time. If the latter is the conceptual basis of a species, then we ask, how is an ancestral species distinguished from any of its descendants? Tokogenetic continuity is maintained in all descendant lineages. How then can we distinguish between a living ancestor and a descendant? Simply put, if tokogenetic continuity is maintained and lineage splitting does not demarcate the end of a lineage, then what does? All that is left is some form of Wheeler and Platnick's Phylogenetic Species Concept: the fixation of a character. However, if the Evolutionary Species Concept lacks a discovery process, it cannot claim a distinction from Wheeler and Platnick's Phylogenetic Species Concept.

Despite their insistence of a lack of a discovery process for species, we find that in practice they defer to the character-based approach in crucial instances, citing the work of Theriot (1992) as a possible example of a surviving ancestor. Theriot (1992) found a plesiomorphically defined cluster appearing in a "four-otomy" with three other apomorphically diagnosed species. According to Wiley and Mayden, the plesiomorphic *Stephanodiscus niagarae* could be considered the ancestral species to the others (*S. reimerii, S. yellowstonensis,* and *S. superiorensis*). But these distinctions were and can only be clearly character based. Apomorphy for *S. niagarae* as a whole would be the only direct character evidence that it had its own evolutionary tendency. That is, an ancestral species cannot be positively found; the hypothesis of an ancestral species becomes an ad hoc explanation for plesiomorphic similarity or must otherwise be the null hypothesis of any analysis that is simply accepted for lack of evidence.

Wiley and Mayden seem to agree with us that distinctions between sexual and asexual species are difficult to make (if not explicitly agreeing with us that the distinction is arbitrary). This explicitly leads them to include asexual species in their version of the Evolutionary Species Concept (contra Frost and Hillis 1990), even though their discussion states that ontologically they are different: that asexual species are like higher taxa and that sexual species are something different. The Evolutionary Species Concept is simply incoherent to us on this point. Moreover, there is a direct epistemological approach to fully asexual species (diagnosis by apomorphy) that is not available to sexual species under the Evolutionary Species Concept.

In short, we believe that the Evolutionary Species Concept as presented by Wiley and Mayden fails in its attempts to defend asexual species and in its argument for existence of living ancestors. At minimum, a species being the smallest independent lineage, the Evolutionary Species Concept seems to require no more or less than any other species concept. In its full extension, considering corollaries, other species concepts deal better with asexual species (either by identifying a logical empiricism, as with our Phylogenetic Species Concept, requiring apomorphy; or by defining them away, as with the Biological Species Concept and Hennigian Species Concept) and ancestors (e.g., the Phylogenetic Species Concept *sensu* Wheeler and Platnick, defining speciation as the acquisition of fixed characters; the Hennigian Species Concept and our Phylogenetic Species Concept, explaining that stem species dissolve at branching points) than does the Evolutionary Species Concept. Overall, the Evolutionary Species Concept is incoherent to us because it fits everything and nothing.

THE PHYLOGENETIC SPECIES CONCEPT *SENSU* WHEELER AND PLATNICK

Proponents of the Wheeler and Platnick version of the Phylogenetic Species Concept argue that it is equally applicable to asexual and sexual species. However, we will show that their concept is clearly not equally applicable to both if such species are to be either the "basic units of . . . organic evolution" or if "under diverse evolutionary processes . . . phylogenetic species represent end products." In the first case, sexual and asexual species simply cannot represent the same kind of unit of evolution. For asexual species, the unit of evolution is the individual. For sexual species, it is some subset of the largest potentially interbreeding array. It cannot be an individual organism. In the second case, asexual species can only be diagnosed by apomorphy in the phylogenetic system because their tokogenetic relationships are hierarchical. Apomorphy is accepted as the best evidence for hierarchical relationships because it is the most parsimonious explanation of fidelic distribution of characters and taxa. There is no such underlying theory for units related through reticulation, whether the units are lineages or organisms.

Wheeler and Platnick's arguments for the use of fixed (but not necessarily apomorphic) characters, the inapplicability of monophyly, the corresponding irrelevance of the concept of apomorphy, and so forth are all built on their assumptions about relationships of sexually reproducing organisms, the need for cladistic analysis to use fixed characters, and the recurrent theme that cladistic analysis must end where reticulation begins. It is self-evident that cladistic analysis cannot correctly identify monophyletic groups where there are none (i.e., where rampant reticulation occurs). However, we fail to see why species must have fixed characters for their discovery unless species are strictly their characters and not lineages. If species are characters, then they are not the units of evolution, only the product. Nothing in what we know about reticulate relationships argues that characters will become fixed; furthermore an independent reticulating unit can be discovered by a combination of traits, rather than by characters. The definition of *fixed* involves layers of assumptions, essentially the same assumptions that go into diagnosing species by a combination of traits: that the population has local boundaries and is well defined and separate from other such units (what Wheeler and co-workers have done is to shift the question of "what is a species?" to "what is a population?"). Their concept does not provide a guide for discovery of relationships on the basis of characters alone. Rather, their definition of character is based on assumptions about relationships (as is the case for all the species concepts discussed in this book, ergo the central empirical problem of species identification).

The modern Hennigian-based view of the relationship of characters and taxa is that monophyletic taxa occur in nature and that they are discovered through the discovery of synapomorphy. They may exist without being discovered. Most simply, it is always more parsimonious to explain similarity as due to common ancestry than not if hierarchical relationships are assumed. Reticulation is a history-destroying process from the perspective of cladistic analysis; reticulation (sex, lateral gene transfer, etc.) is one explanation among many (reversal, convergence, misinterpretation, etc.) of nonfidelic or homoplastic character distributions. Thus, there is no expectation or requirement for discovery of fixed characters to be taken as evidence for lack of reticulation, or conversely, for lack of discovery of fixed characters to be taken as evidence for reticulation. As Wheeler and Platnick have said about Hennig's definition of monophyly, the principles of cladistics are applicable only to hierarchical patterns of descent. Such principles cannot be used to argue for or against the need/expectation of fixed characters where there is no such pattern.

Unlike that of sexual species, the history of individuals of asexual species (whether unicells reproducing by division or parthenogens) can be represented by a branching diagram. The terms used by Hennig to describe characters (plesiomorphy, apomorphy) are perfectly applicable to asexual species, and the ontological status of purely asexual species can only be equivalent to that of higher taxa. Thus, the concept of Wheeler and Platnick cannot be applied to asexual species and still be said to identify unique products of evolution any more than a combination of plesiomorphic traits diagnoses a monophyletic Reptilia. Rather than including asexuals in a discussion of the ontological status of species, Wheeler and Platnick have just grafted them onto arguments they began based on sexually reproducing species. That is because their arguments are based not on a lineage concept, but on the distinction between reticulation and branching and on the need of higher-level cladistic analysis to have fixed characters, distinctions not appropriate to asexual species because they lack reticulation and by definition have fixed characters.

There is no ontological reason whatsoever to assume that nonapomorphic characters identify a lineage of any sort in asexual species: asexual tokogeny is hierarchical (at least in organisms that divide). However, it is a bit more difficult to see hierarchy in parthenogenetic creatures. In asexual amoebae, the individual cell is the lineage, much like Hennig's species. The life of the cell is demarcated by cyto/karyokinetic events (speciation). In parthenogenetic species, the ancestor survives. However, the relationship between individuals is still branching, and the synchronic cut method of monophyly still applies. This distinction is worth emphasizing: not all asexual species are amoeboid. The parthenogenetic case also points out the importance of the metaphyly ontology (Donoghue 1985; Mishler and Brandon 1987). A parent may be neither paraphyletic nor monophyletic with respect to its offspring, resolving Wiley and Mayden's paradox on this issue.

Wheeler and Platnick argue that it is an advantage of their Phylogenetic Species Concept that it is independent of cladistic analysis. This point, however, is never explained. There is no ontological reason why this is an advantage. Conversely, just as Hennig reasoned that cladistic analysis is necessary for identifying higher monophyletic taxa, it may be that such reasoning is necessary for identifying species or species taxa. It would certainly be easier to have a species concept that did not require cladistic analysis, but it would be also be easier to have one that excluded molecular genetic analysis or tedious dissections. As we have shown, cladistic analysis is absolutely necessary for determination of natural lineages of any level in asexuals.

Using paraphyletic units as terminals can mislead a cladistic analysis (insofar as relevant characters are not monomorphic for the paraphyletic terminal). But that is true for all concepts and at all levels of analysis. Furthermore, there is no reason why the basal units of cladistic analysis should be a priori identified as species. Cladistic analysis requires only the assumption that cladistic relationships exist among terminals and that characters be heritable and independent. There is no assumption that the terminals should be monophyletic, natural, formally named species. As long as the terminals are themselves open to testing, cladistic analysis does not require circular reasoning. A species concept that requires cladistic analysis of its terminal units is no more circular than one that requires cladistic analysis of its more inclusive units.

We also disagree with Wheeler and Platnick on the ranking issue. One simply cannot and should not formally name all the smallest monophyletic groups one finds. There need be no fundamental or basic taxonomic unit; in fact, one does not exist. As a result of evolution, there are nested monophyletic groups. The decision about which monophyletic groups are taxa comes later and includes practical human concerns. Taxonomy is clearly a human endeavor, a legislated set of conventions, so formally named taxa can all be regarded as practical decisions by taxonomists. We are and should be realists, but given indefinitely many real phylogenetic groups, the taxonomist must decide which to name based on their perceived importance (i.e., phylogenetic groups about which one has something important to say, processwise, or perhaps groups that have particularly strong support). We want to make a clear distinction between the ontological status of monophyletic groups (which are real entities for which we may or may not have epistemological evidence of their existence, i.e., apomorphies) and the ontological status of taxon names (which are human constructs ontologically even though we use the same kind of epistemological evidence to document them). There are thus two ontological issues at stake (that we and others have sometimes confused): the ontological status of monophyletic groups and the ontological status of taxa. There is an inclusion relationship between the two because all taxa are monophyletic groups, but the reverse is not true, indicating that a different ontology applies.

We feel that the species concepts of Wiley and Mayden, Meier and Willmann, and Mayr can be rejected out of hand for the reasons detailed above. They are fatally based on wrong ontologies having to do with interbreeding groups. We are more sympathetic to the approach held by Wheeler and Platnick; we share a similar ontology (even though theirs is not fully specified) and epistemology.

However, our differences with Wheeler and Platnick can be summarized under two headings: grouping and ranking. We disagree with them in real (but probably relatively minor) ways about how to put organisms into groups using phylogenetic characters. This difference has more to do with our respective ontologies than with empirical activities. When faced with a case study, both we and Wheeler and Platnick would look for characters and conduct as detailed an analysis as is feasible. A more basic disagreement is that we treat the species level like all other levels ontologically and empirically. We name only selected clades that we have reason to discuss (i.e., those with solid support, ecologically important apomorphies, etc.). Although there is presumably a smallest monophyletic group, ontologically speaking, in any particular case (i.e., the level at which processes causing divergence have prodominated over processes causing reticulation, for the time being at least; Horvath 1997), we do not want to automatically call this smallest monophyletic group a species. We reserve the right not to formally name all known lineages (e.g., in cases where the smallest known lineage is really tiny and insignificant, doubtfully stable, etc.). The Linnaean system is legislated by humans for their own purposes of communication. It is alright for it to be subjective if by that we mean naming only those lineages (at whatever level) we have reason to talk about.

10

A Critique from the Wheeler and Platnick Phylogenetic Species Concept Perspective: Problems with Alternative Concepts of Species

Quentin D. Wheeler and Norman I. Platnick

The goal for developing a Phylogenetic Species Concept is to support the aims of phylogenetic systematics: to discover elements for reconstruction of phylogenetic history, distinguish among kinds of organisms, describe and predictively classify the earth's biological diversity, and permit the study of evolution and comparison of clades. In order for a species concept to be phylogenetic, it need only accurately identify those elements among which evidence of shared history may be unambiguously retrieved; cladistic analyses of subdivisions of phylogenetic species need not, and indeed should not, be possible.

We will discuss alternatives to our Phylogenetic Species Concept presented in preceding chapters and identify the properties that make them inconsistent with phylogenetic theory or practice.

The Biological Species Concept

It is unclear in Mayr's chapter just how he views our (phylogenetic) concept of species. He provides criticism of aspects of our concept under both typological and phylogenetic species and seems to confuse our concept with that based on autapomorphy (i.e., Mishler and Theriot's Phylogenetic Species Concept).

Mayr repeats his long-standing complaint about authors who confuse the concept of species (i.e., species as a formal Linnaean category) with the species as a taxon. Mayr makes, whether intentionally or carelessly, a similar mistake when he (at least implicitly) equates morphological and typological species. The use of morphological evidence does not imply typological thinking. Nor does the typification of species imply that all individuals of a species are either invariant or identical to a holotype. True typology disappeared from mainstream taxonomy long ago. Most contemporary authors who find morphologically based species compelling base their concepts on theories that allow for the kinds of infraspecific genetic variability that contributed to the demise of typology.

Mayr refers to intrinsic morphological differences among species to draw an equation between morphological species and typological ones. What is inherent is not some Aristotelian essence or arbitrarily chosen structure, but rather shared genetic information, relevant to the retrieval of phylogenetic history, that is most parsimoniously interpreted as the result of common ancestry rather than parallel evolutionary events.

Mayr makes repeated claims that phylogenetic (or at least his "typological") species are based on a degree of phenetic difference. For our Phylogenetic Species Concept, this is simply false. The accusation is ironic because recognition of subspecies—a practice endorsed by Mayr (1963)—is based on arbitrarily selected degrees of difference among populations. Phylogenetic species are based on constantly distributed characters shared by all constituent individuals. Such characters need not be apomorphic, as for example in the case of an ancestral species, and certainly need not be invariant (see Platnick 1979). Just as phenotypic differences in the neck of a mouse and giraffe do not contradict the constancy of the character *backbone* for the higher taxon Vertebrata, so variations among individuals within a single species do not necessarily negate the existence of a constantly distributed attribute. However, the constancy criterion does preclude the kind of subjectivity associated with reliance upon degree of resemblance. Such subjectivity is far more evident in applications of the Biological Species Concept, where peripheral populations are arbitrarily treated as subspecies, closely related species, or just another population, often based on little more than the degree of phenetic resemblance. Where they exist, well-defined, diagnosable "subspecies" should simply be called species.

Morphological (and, we presume, phylogenetic) species are viewed by Mayr as classes made recognizable by defining characters. Species have some of the properties that philosophers have traditionally attributed to classes, as well as some of the properties that philosophers have traditionally attributed to individuals. Discrepancies between species and concepts of classes and individuals are problems for philosophers, not biologists. Phylogenetic species, like other testable scientific hypotheses, should be based on and make predictions about observable (character) evidence. The Biological Species Concept is said by Mayr to be based on the observed presence or absence of interbreeding, and yet such observations are rarely attempted and far less often attained. For a very large number of species, no such observations are easily or affordably obtained, making applications of the Biological Species Concept highly speculative. As increasing numbers of species become extinct in the course of the biodiversity crisis (Wilson 1985, 1992), many species will be known only from museum specimens. In order to determine whether such material represents previously known or unknown species and in order to analyze how those species relate to other living or extinct species, a concept of species will be required that can be used in practice without experimental observation of interbreeding.

Also, if we are to succeed in learning about the diversification of life on Earth, we must be able to compare clades of distantly related kinds of organisms. Are there more species of flowering plants or flies? Do asexually reproducing clades generally produce more species than sexual ones over comparable periods of time? Such questions can be answered only if species in one clade denote something comparable with species in another. The Biological Species Concept by admission is inapplicable to asexual forms.

Mayr asserts that ecologists and ethologists working with local populations of species are not affected by whether taxonomists decide that some remote population is or is not conspecific with their subjects of study. To the extent that this is the case, it is a harsh criticism of ecology and ethology in that they are conducting their work outside of an evolutionary context. In order to understand what is happening at the local level in regard to evolution, it is necessary to make comparisons broadly, and ultimately to interpret observations in the context of a phylogeny of all related species regardless of where they live. Similarly, Mayr says that it is irrelevant for most biological investigations whether isolated populations with differences equal to those generally accepted as species-level differences are regarded as subspecies or full species. Such nonevolutionary arguments unnecessarily constrain most kinds of experimental biology to a very limited descriptive role devoid of demonstrable evolutionary relevance.

We do not deny that Mayr's biological species are often easily recognized by the local naturalist. In fact, it is this provincial visibility of distinct kinds that first suggested to biologists that species exist. A species concept that meets the general needs of biology to understand the origin and diversification of species and clades, however, must be applicable across the full geographic and temporal ranges occupied by kinds of organisms. Criteria must exist to recognize species in the fossil record and in distant lands, and such criteria must themselves be open to potential falsification. Interbreeding, much less the propensity to interbreed, is not applicable across wide spans of geologic time or geographic space; therefore, the Biological Species Concept is theoretically and practically limited.

Mayr erroneously claims that what constitutes a diagnosable difference among phylogenetic species is totally subjective. Hypotheses about homology and character distribution are fully testable based on their predicted occurrence. The claim that a particular attribute is a defining character is open to potential falsification, whereas the subjective decision to recognize allopatric populations as either species or subspecies under the Biological Species Concept requires the kind of subjectivity that Mayr (and we) find objectionable.

According to Mayr, morphological species concepts (presumably including our Phylogenetic Species Concept) do not share the biological significance he ascribes to his species concept. To the contrary, our Phylogenetic Species Concept recognizes the distinction between tokogenetic and phylogenetic systems in sexually reproductive organisms (cf. Hennig 1966), makes testable hypotheses about the observable evidence of phylogeny, and identifies the products of shared evolutionary history among which evidence for that history exists. In an evolutionary sense, these are fundamental, essential, and biologically relevant units that make possible the fruitful comparison of different kinds of life forms. Neither the smallest diagnosable unit nor that evidence that makes species diagnosably distinct are arbitrary, as Mayr claims (cf. Nelson and Platnick 1981; Nixon and Wheeler 1992a).

Mayr says that similarities revealed by molecular geneticists point to similar propensities in clades and provide a logical basis for expecting parallelophyly of identical characters repeatedly. Sorting out whether similarities are indicative of common ancestry or homoplasy is a complex undertaking, ultimately dependent on the parsimony criterion (Farris 1983; Carpenter and Nixon 1994). Contrary to Mayr's conclusion, our Phylogenetic Species Concept is as fully applicable to the field biologist as to the museum taxonomist.

Mayr is more deterministic in his view of species than we are: "The species is a device for the protection of harmonious, well-integrated genotypes." The number of factors contributing to the origin of species is simply too great to support such a sweeping generalization, including a broad range of biological and abiotic forces acting singly and in varied combinations. Some, perhaps many, species result from isolation imposed on them whether or not they are well integrated genetically and whether or not they need protection. This fact is implicitly recognized by Mayr, who states that isolated populations may only share a propensity for interbreeding.

Finally, proponents of the Biological Species Concept seem generally concerned that our Phylogenetic Species Concept would lead inevitably to a drastic increase in the number of species. Although our Phylogenetic Species Concept will result in a sharp increase in the number of species for a few taxa such as Aves (J. Cracraft, personal communication), the impact will be negligible for most major clades such as Arthropoda. Furthermore, a fundamental goal of all species concepts is to discover how many kinds of organisms exist. Underestimating their numbers simply to avoid recognizing what is arbitrarily deemed as too many species seems biologically indefensible.

The Hennigian Species Concept

Hennig's concept of species was essentially a version of the Biological Species Concept, modified in an effort to circumvent problems incurred with a paleontological perspective. Interbreeding criteria found useful at one time horizon dissolve as one considers populations extending backward through geologic time. Because all descendant populations within sexually reproducing species presumably interbred at some remote time (Rosen 1979), such an assertion is hardly surprising. In fact, similar problems exist for concepts based on interbreeding criteria even among synchronous populations. Intermittent, low-level gene flow among partially isolated populations renders decisions regarding the status of biological species necessarily subjective.

According to the Hennigian Species Concept, stem species do not survive speciation events; hence an ancestral species ceases to exist at the time of origin of its daughter species, even when one of those daughter species is identical to that ancestor. It is through this dissolution of the common ancestral species that Meier and Willmann define Hennigian species. Few scientists have taken Hennig's termination of stem species literally. Rather, Hennig's view has been seen as a methodological convention since recognition of similar ancestor-descendant species may be problematic. Given this convention, all species being analyzed may be treated as terminals on the cladogram, regardless of whether or not any ancestral species are actually included. The result, of course, is a phylogenetic pattern projecting the relative relationship among these terms. Any ideas about stem species or the actual identity of ancestor species go beyond such analysis and become speculative because the ancestor is not expected to exhibit any apomorphies not shared by its descendants (Nelson and Platnick 1981).

Meier and Willmann's definition is theoretically objectionable in its summarial rejection of the possible survival of an ancestral species beyond a speciation event. Because their ancestral and daughter species can be identical in all respects, there is no justification for conceiving of them as different species. Moreover, it is indefensible in practice because it calls for belief in historical events for which no known unequivocal source of evidence exists. The origin of a reproductive gap, such as the isolation of two allopatric populations due to the origin of some geophysical obstacle, does not necessarily correspond with character transformation or, by extension, with speciation. Taken literally, the isolation of a pair of white-tailed deer in the basement of the American Museum would result not only in their being reproductively isolated and thereby a new species, but in the even more remarkable extinction of the ancestral species of deer. No character transformation (read *evidence*) is called for by Meier and Willmann, just knowledge of the origin of a reproductive gap that may or may not ever result in any change (i.e., speciation in the usual sense). Whether the origin of such gaps could be observed as they happen is questionable; that they could be recognized at all in the remote geological past is no more than wishful thinking.

Meier and Willmann claim that Hennigian species are recognized relative to their sister species rather than any other species. This, however, is problematic in both theory and practice. Meier and Willmann define Hennigian species, after all, on the basis of presumed dissolution of the ancestor, not properties of sister species. In the context of phylogenetic studies, sister-group relationships are relative statements that may rarely be taken literally. If the paleontological view that vast numbers of species have become extinct in the past is correct (see Mayr 1982; Novacek and Wheeler 1992), then actual sister species are only rarely included in an analysis. Because the number of missing species between those species included in an analysis varies widely, Meier and

Willmann's insistence on comparisons with sister species must be relaxed in virtually every case. Meier and Willmann ask for results of a cladistic analysis in order to recognize sister species, even though species must be recognized prior to such analysis.

Meier and Willmann are necessarily vague about when a reproductive gap becomes a species gap. Some gaps open, only to subsequently close in the face of renewed gene flow. How is a biologist to predict which reproductively isolated populations are isolated permanently and which may ultimately merge again into a single species? Because their concept is divorced from the character evidence that marks such permanent divergence, it is left with subjective measures of genetic isolation and reproductive gaps that are imprecise in theory and beyond the bounds of empirical observations.

Meier and Willmann cite four objections to morphologically based species concepts, none of which are valid with regard to our Phylogenetic Species Concept. First, morphological evidence is said to vary continuously so as to undercut any effort to circumscribe species. Where attributes are inconstantly distributed, it is likely that we have failed to distinguish between traits and characters (Nixon and Wheeler 1990, 1992a) or that we have not recognized the relationship between an original attribute and some of its modified subsequent forms (Platnick 1979). The actual problems here are familiar ones related to homology, not deterrents to successful recognition of species. Second, morphological differences are presumed to disappear as one traces them backward through geologic time. Again, those attributes providing evidence of common ancestry are only those transformed in a common ancestor and passed on to its descendant species in its original or some modified form; as such, characters do not dissolve through time because their first occurrence is predicated on constant distribution in the ancestral species (Nixon and Wheeler 1992a). Third, species boundaries are said to depend on which morphological features are selected. Characters may be selected, of course, on the basis of their observed distributions so that they are not arbitrary. Furthermore, such characters are open to falsification because their distribution among populations is clearly predicted. Finally, although problems associated with differences among clustering algorithms are valid complaints against phenetic approaches to this problem, they are simply inapplicable to our character-based Phylogenetic Species Concept.

Meier and Willmann imply that there is something wrong with recognizing species on the basis of plesiomorphies. This argument appeals to phylogeneticists because of the necessity of synapomorphy for grouping species together into monophyletic clades. However, such logic does not apply at the level of species, where the objective is to recognize elements among which phylogenetic history may be retrieved but within which it may not. Given an ancestral population that is dimorphic for some character, the fixation of each alternative state in daughter species can define two new species equally efficiently. Considering the presence of each attribute in the common ancestor, on what basis is one fixed state more apomorphic than the other? They are each similarly removed from the original polymorphic condition and are thereby equally derived. Meier and Willmann specifically warn that the application of our Phylogenetic Species Concept to asexual lineages may result in species based entirely on plesiomorphy. The same may eventuate for sexually reproductive organisms. In neither instance, however, is this a problem. The stem species for any clade has only those apomorphies shared also by all its descendants. It has no autapomorphies of its own. Does this mean that ancestral species did not exist or that they may not be included in a cladistic analysis when they are available? They did exist and may be included in an analysis.

It seems as though Meier and Willmann wish to take literally the lines that systematists draw connecting species in a cladogram, ascribing to them biological properties that transcend observable evidence such as recognizing as a species only one part of a stem. In any analysis, we are constrained by what we can see and interpret. Species are hypothesized to be sisters only relative to whatever additional species are known and included in the study. Extinct populations, even when they are known to us, are not unequivocally assignable to particular temporal spots along an ancestral stem. When do ancestral species cease to exist or daughter species begin? A meaningful answer is based on character transformation, the only affirmation of either speciation or cladogenesis.

Meier and Willmann say that asexual lineages are hierarchic and that our Phylogenetic Species Concept creates nonmonophyletic groups that "obscure the phylogenetic structure within uniparental taxa." Perhaps such relationships would be obscured by alternative concepts of species, but clearly not by our Phylogenetic Species Concept. Where such phylogenetic structure can be shown within a phylogenetic species, we have simply underestimated the number of phylogenetic species present in the sample.

Meier and Willmann assert that phyletic speciation becomes rampant when our Phylogenetic Species Concept is applied. The proximal aims of systematic biology are not to determine which of the many processes that shape evolution were operative in a particular instance, but to recognize the end-products of that history and to analyze the relationships among them. Species may be objectively defined and recognized relative to the only conceivable scientific evidence for either speciation or phylogeny: characters. In this context, various modes of speciation are indistinguishable. In general, we are concerned about any scientific concept that is imposed in the absence of testable, observable evidence. Meier and Willmann's insistence on the nonconspecificity of mother and daughter species that are identical in every respect is simply a belief divorced from evidence.

Curiously, Meier and Willmann would not recognize more than one species within a phyletic lineage in which major character transformations had occurred, yet would insist on the separate identities of mother and daughter species that were never isolated and had not diverged in even a single respect.

The Phylogenetic Species Concept *sensu* Mishler and Theriot

Mishler and Theriot find it odd that the Phylogenetic Species Concept proposed by Eldredge and Cracraft (1980), Cracraft (1983), Nelson and Platnick (1981), and Nixon and Wheeler (1990, 1992a) was "explicitly not based on synapomorphy." On the other hand, we find it peculiar that Mishler and Theriot believe that in order for a species concept to be phylogenetic, it is insufficient for it to be fully compatible with phylogenetic theory. Instead, they would insist that species be the result of phylogenetic analyses of infraspecific units. Such a dependency of species on cladistic analysis is unnecessary and problematic.

Phylogenetic history is retrievable above the level of species, in part because of the emergence of hierarchical relations marked by character transformations. Without character transformation in sexually reproductive organisms, there are only reticulate patterns of tokogenetic relationships.

It was not accidental that Hennig discussed his concept of monophyly above the level of species. Given his distinction between tokogeny and phylogeny (Hennig 1966), it was logical that phylogenetic history would be retrievable above the level of species and not below. For phylogeneticists, then, the species concept plays a key role in recognizing the least inclusive terms among which cladistic relationships may be analyzed. Making species dependent on cladistic analysis implies that Hennig's tokogeny-phylogeny distinction was not valid and that *monophyly* has a meaning other than Hennig's.

Taking Mishler and Theriot's definition seriously, ancestral species cannot exist because they are plesiomorphic in all respects, relative to their own descendants, and hence not analogous to monophyletic clades. They try to escape this biologically uncomfortable by-product of their species concept by defining species only at a single time horizon. This is a curious property for elements ostensibly recognized for the expressed purpose of studying history. Perhaps the use of the term *monophyly* to describe non-hierarchical systems is a misapplication that, if followed, would undermine the logical clarity given that term by Hennig.

For Mishler and Theriot, species are artificial constructs in the minds of biologists. Given their concept of species, based on inappropriate application of the notion of monophyly within populations, this position is understandable. However, traits can give way through character transformation (essentially, the removal of ancestral polymorphism via extinction events) to characters. Thus, a level exists for sexually reproductive kinds of organisms below which tokogenetic relationships make unequivocal retrieval of historical relationships impossible and above which historical patterns of common ancestry among species may be retrieved and critically tested. Such a level is dictated not by the minds of biologists, but by the distribution of characters among populations.

Ironically, Misher and Theriot deny the existence of species at the same time that they acknowledge the existence of demes. Local populations in reality share nothing more than geographic proximity. Actual interbreeding does not take place among all members of any sizeable deme, and immigration, emigration, birth, and death of individuals keep demes in constant flux. Demes do not exist for more than fleeting moments outside the minds of biologists, and they have little more biological reality than Cleisthenes' townships of Attica, for which they are named.

Inexplicably, Mishler and Theriot criticize our concept for an overemphasis on epistemology at the expense of ontology while they advocate the blind use of cladistic analyses at levels that violate everything we have learned about shifting gene frequencies within populations. All that we know about how tokogenetic systems function suggests that relationships among their parts are not predictably hierarchical. All that we know about characters suggests that their transformation provides the only conceivable evidence with which to reconstruct phylogeny. Ontologically, cladistic analyses are justified by a reasonable expectation for hierarchical relations among species, groups of species, and nested sets of characters.

Mishler and Theriot claim that cladistic relations exist down to the level of individual organisms within asexually reproductive creatures, with the clear implication that our concept must either be arbitrarily applied or result in the recognition of individuals as species. The clever reader no doubt will have noticed that our concept refers to asexual lineages, meaning lines of mother-daughter individuals (often very large numbers of them) that share particular characters. Until two or more individuals exist that are identical in those characters, no lineage exists and no phylogenetic species exists.

Were reticulation and parallelism the dominant pattern, there would be no ontological justification for expecting evolutionary history to result in hierarchical relationships. Taking Mishler and Theriot's arguments about commonality of reticulation to their logical extreme, one would be inclined to conclude that retrieval of historical patterns is impossible at higher levels rather than concluding that monophyly as a concept applies within sexual populations.

Woodger (1937) gave a formal definition of hierarchies that included specific attributes: that is, a *unique beginner* (an element that itself is not divisible into smaller elements), a single direction of change, and a progression from one to many. Neither individual organisms nor populations of sexually reproducing organisms fulfill Woodger's criteria. For genealogies of such sexual forms within populations, no unique beginner exists because both a male and a female parent are required to produce a new individual. In the case of populations, the beginner is not elemental, because populations may be subdivided into similar, if smaller, populations. And there is no unidirectional progression because gene frequencies are reversible. Monophyletic clades and lineages of asexual species, on the other hand, are fully hierarchical. It is the existence of such hierarchies that accounts for retrievable nested sets of characters (see Platnick 1979; Nixon and Wheeler 1992a). Therefore, a *fundamental level* does exist, at least in the context of sexually reproductive organisms, that can equate to species.

Mishler and Theriot admit that "it makes no sense to apply a species concept that requires prior specific knowledge of processes," yet their decision to rank groups of individuals as species "might well include ecological criteria or the presence of breeding barriers in particular cases." This would seem to be unnecessary for phylogenetic theory, practice, or history (cf. Nelson and Platnick 1981).

Certainly, cladistic analyses have revealed that homoplasy is in no short supply and that reticulation and parallelism exist in nature. Tokogenetic and phylogenetic systems differ, however, with regard to whether it is reasonable to expect a hierarchical pattern to obtain.

The Evolutionary Species Concept

Wiley and Mayden make many points that, when taken out of context, seem perfectly agreeable. For example, monophyly does not apply to individual and population levels, species are lineages, and species that are divisible into diagnosable geographic parts underestimate the number of species present. We agree also that their Evolutionary Species Concept is consistent with most of phylogenetic theory. In fact, one of our greatest problems with the Evolutionary Species Concept is the fact that it is similarly compatible with virtually any pattern or process in evolution. Saying that lineages exist or that they have histories or tendencies or fates are no more than vague assertions that evolution exists. We are willing to stipulate that evolution has occurred and, contrary to many of our critics, find within that seemingly simple assumption sufficient ontological justification for phylogenetic species.

Wiley and Mayden claim that Hennig (1966:32) repudiated morphologically based definitions of species as typological. To the contrary, Hennig (1966:32) underscored the fact that holomorphological similarities are expressions of shared genetic information and thus supported the utility of morphological comparisons of semaphoronts and, by extension, species.

Wiley and Mayden contend that monophyletic groups do not exist because of synapomorphies. We agree with what we take them to mean: that merely grouping together individuals that have some attribute in common does not make that group one with evolutionary significance. However, because character transformation and speciation are inseparable outcomes of one and the same process—the extinction of ancestral polymorphism—then monophyletic groups really do exist because of the transformation of one or more characters in the original ancestor of the clade. Our Phylogenetic Species Concept is said by Wiley and Mayden to be operational and, as a consequence, to suffer from four significant problems. First, because characters are neither sufficient nor necessary to define taxa, a species concept that incorporates characters into its definition is insufficient. This supposes that epistemological aspects of our Phylogenetic Species Concept are divorced from ontological considerations; we shall address this recurring accusation under general conclusions and will say here only that the inclusion of characters does not deny ontology. Second, a definition that embodies a discovery method confuses "what things are" with "how things are discovered." Neither we nor our Phylogenetic Species Concept are confused in this regard. Again, we refer to the discussion of ontology that follows. Third, operationalism is said to sacrifice accuracy for precision unless processes are so well known as to make accuracy and precision synonymous. Operations of our Phylogenetic Species Concept, however, are calculated to precisely and accurately recognize the products of evolutionary history, regardless of the particular processes that may have brought that history about. If we are wrong to assume that hierarchical relations within tokogenetic systems are equivocal at best, then we suppose that accuracy might be at risk. Such has not been demonstrated, however, and specific applications of such assumptions are open to falsification (i.e., characters may be shown to be distributed as traits or vice versa). Finally, operationalism is said to not be equivalent to testability. This is certainly so, although our Phylogenetic Species Concept is explicitly testable.

"The evolution of organisms receives its very special character—which is also decisive in the tasks of systematics—from the existence of species" (Hennig 1966:197).

Species need not be the artifice of human thought, although Mishler and Theriot so conclude in light of their species concept. Species have long been recognized as self-perpetuating kinds of organisms (e.g., Buffon 1749). We, like Mayr, Meier and Willmann, and Wiley and Mayden, believe that species exist and that they are discoverable. Species, for us, are not merely another ranking decision in an infinite continuum from molecules to clades.

What are the goals of a species concept? For us, they include the following:

1. To recognize the kinds and numbers of distinct, self-perpetuating organisms on Earth, past and present.
2. To identify the end-products of diverse evolutionary processes.
3. To discover the elements of phylogenetic analysis, that is, those groups of organisms among which there is a retrievable common history and which may not be divided into less inclusive units for which the same is true.
4. To determine the least inclusive units usefully accorded formal recognition in a Linnaean classification, consistent with the goals of communicating and predicting the distribution of characters among organisms. In this sense, species names occupy a special place as the expression of taxonomists' least inclusive hypotheses preliminary to cladistic analyses.

The dismissal of asexual organisms from species concepts on the basis that they originate from a different evolutionary process is illogical because a vast array of evolutionary processes act individually and in concert upon sexually reproducing species. Were this true, then some kind of pluralistic view of species might be required, and no comparison of species or their numbers among distantly related taxa would be valid. We are not willing to concede that such is the case unless and until all conceivable concepts of species have been fully explored and rejected. Our Phylogenetic Species Concept appears in theory to address this problem by recognizing all end products of evolution (whether agamotaxic lineages or groups of populations) so long as they are marked by the transformation of one or more characters. We believe that this treatment of kinds of asexual organisms is consistent with our goals for recognition of species.

What the alternative concepts of species discussed in this volume have in common is a greater or lesser dependence on assumptions about evolutionary processes. This stands in sharp contrast to our Phylogenetic Species Concept, which depends on the distribution of observable characters. This evidentiary basis for the phylogenetic species is fully consistent with phylogenetic systematics.

It should be remembered that cladistic analyses seek to discover patterns of relationship shared among species. It is necessary and logical to recognize species prior to such an analysis. This, in turn, necessitates criteria for their recognition and a species concept that does not depend on cladistic analysis. Thus, any concept that demands supposed monophyly for species or knowledge of sister species is seriously misguided. Outside of cladistic analysis, knowledge of monophyly is unattainable. There are compelling theoretical, practical, and historical reasons to believe that the use of the term *monophyly* below the species level is inappropriate and misleading (see Nixon and Wheeler 1990). Even if the term *monophyly* could be applied within tokogenetic systems, there is no ontological basis for expecting observed patterns to be monophyletic. In the case of asexual systems, one simply divides clades until the most complete hierarchy of clonal relations has been revealed.

Just as the history of phylogenetic systematics has witnessed a refinement of theories and methods that permit the recovery of historical patterns independent of unnecessary assumptions about specific evolutionary processes, the development of a phylogenetic concept of species has gradually replaced similar process assumptions with factual, objective criteria to sort out those groups of organisms among which unambiguous evidence of history exists. Each competing concept, for one or more reasons, falls short of this objective and adopts evolutionary assumptions that are either indefensible or simply unnecessary. Our Phylogenetic Species Concept avoids such problems and simply and efficiently meets the full range of goals for a species concept within contemporary systematic biology.

A recurring attack upon our Phylogenetic Species Concept is founded on the belief that the mention of characters in our definition renders our Phylogenetic Species Concept devoid of ontological content. Although this explicit reference to characters emphasizes the epistemological practicality of our Phylogenetic Species Concept, it is also an explicit reference to a body of theory that makes the term *character* meaningful in a precise and specific way. Without the underlying assumptions that evolution has occurred and that this has been accompanied by a history of character transformations, there would be no ontological basis for our Phylogenetic Species Concept.

Although all the alternative concepts of species discussed by other authors in this volume strive to meet the need for elements of phylogeny, each suffers from the unnecessary inclusion of assumptions about process, reliance on cladistic analysis or knowledge, or other problems of theory or practice that can be avoided by adoption of our Phylogenetic Species Concept.

11

A Critique from the Evolutionary Species Concept Perspective

E. O. Wiley and Richard L. Mayden

Perhaps some of the differences among systematists regarding their views of the nature of species are due to their different views of how concepts function in the phylogenetic system and science in general. To Wheeler and Platnick, a concept seems to function as an operational device allowing one to discover something. To Mayr, a concept seems to function as a biological description of how entities should behave if they are distinct things. To Mishler and Theriot, the world is too complicated to allow for a monistic concept *sensu* Kluge (1990).

We believe that concepts should serve as bridges between pattern and process, links between the empirical and theoretical worlds that serve as checks on our collective wisdom or lack thereof. Concepts that prohibit nothing cannot allow us to test process theories. Well-formulated concepts may be abandoned because they fail to provide an adequate bridge as empirical data increase and affect how we perceive processes thought to occur in nature. This series of papers provides examples of both phenomena. Wheeler and Platnick state that their species concept is compatible with any hypothesized mode of speciation. If true, then patterns of species relationship cannot be used to discriminate between different modes of speciation. We do not find this useful. Forms of the Biological Species Concept are characterized primarily by reproductive closure either in the "soft" sense of Mayr's Biological Species Concept or the "hard" sense of the hyperbiological species concept of Meier and Willmann. These seem to be perfectly good concepts. The problem is that these concepts do not always seem to fit the empirical pattern provided by phylogenetic analysis. Neither is the Evolutionary Species Concept immune from criticism. If one compares the formulation by Wiley (1978) with our current formulation in this volume, one can note that evolutionary species are no longer characterized as lineages composed of populations. This is an accommodation to criticisms that asexual entities do not form populations but clone vectors. Thus, concepts must adapt or be abandoned as we gain greater insight into the evolutionary process.

We believe that this series of papers points to a more general struggle that has occurred in many branches of science. First, do we treat concepts of species as real under the general paradigm of logical realism, or do we treat them as nominal? We choose to treat them as real. Put simply, we view the Evolutionary Species Concept as a universal concept requiring discovery of particular examples (*Homo sapiens, Fundulus lineolatus,* etc.). Furthermore, we submit that the Evolutionary Species Concept is the most general concept available. Second, do we treat our definitions as analytical statements or nonanalytical statements? If we treat them as analytical statements, then the theoretical term (e.g., *species*) is coupled with a statement to form a concept (definition) that has testable consequences. Various operations can then be performed to see if these consequences obtain for a particular case. For example: "An atom is the smallest unit of matter that retains the chemical properties of the class of elements to which it belongs" is taken as an analytical statement despite the fact that no particular property is specified and no particular operation is specified to find the property. This is not bad. In fact, it is good because the statement allows us to discover new kinds of atoms. There are plenty of ways one might go about investigating the chemical properties of a molecule to see whether it is composed of atoms with similar or different chemical properties. These operations do not have to be built into the definition because the effects one might find using the operations are logical consequences of the analytical statement itself. Operational definitions are needed only when working with nonanalytical statements that can be defined only by the operations themselves. The only things that will be discovered are those allowed by the operations performed. This would be fine only if the operations ensured that all relevant entities and only relevant entities would, in fact, be discovered. We see no presently articulated operational definition of species that performs this task.

We have two choices. Do we proceed from theoretical term (cf. *species*) to analytical statement (cf. "is a lineage . . .") and then to operations that prove or disprove that a particular hypothesis is an example of the theoretical term? Or do we proceed from theoretical term directly to operations because an analytical statement is impossible? To put it simply, is the statement "a species is a lineage evolving independently of other lineages" an analytical or a nonanalytical statement? It must be analytical because the concept of an independently evolving lineage is linked to all sorts of testable phenomena in the same manner as the concept of the smallest unit of matter with its own chemical properties is linked a to all sorts of testable phenomena. Two examples will suffice. The fact that an investigator could diagnose a newly discovered group of populations would be evidence of lineage independence. In the same manner, the statement "species are closed reproductive communities" is an analytical statement, and we have no doubt that many independently evolving lineages have this quality. We submit that the need for operational definitions of terms is a myth. It is the analytical statement and not the operational definition that we should be seeking. That analytical statement is operational because it contains elements that are testable through empirical investigation. Perforce, the more general the concept, the more general the analytical statement. This is no reflection on the concept except to say that it is more general.

Phylogenetic Species *sensu* Wheeler and Platnick

Wheeler and Platnick equate speciation with the fixation of an apomorphic character. If we assume that species are what speciation yields, we might conclude that this concept is just the apomorphic species concept of Rosen (1979) and the monophyletic species concept of Donoghue (1985). Yet, Wheeler and Platnick explicitly reject the notion that the concept of monophyly applies to species. Apparently, species need not have apomorphies for three reasons. First, we might discover and name our species before phylogenetic analysis reveals the derived characters. Second, even after the phylogenetic analysis, we might not be able to polarize two homologous characters if the ancestor was polymorphic for these characters. Third, the apomorphy of an ancestor becomes a plesiomorphy upon speciation, providing for the synapomorphies of higher taxa. Of these three reasons, only the third is legitimate. Donoghue (1985) might argue that initial hypotheses of species identities can be set aside after the phylogenetic analysis, bringing binominals in line with monophyly in the same manner as initial hypotheses of monophyly for higher taxa may be changed with subsequent analysis. The second is biologically incorrect. All apomorphies must go through a polymorphic stage where they coexist with plesiomorphies. If speciation resolves itself into the situation where one species retains the plesiomorphic character while another species is fixed for the apomorphic character, then we can only assume that the plesiomorphic character is, in fact, plesiomorphic by some relevant criterion such as outgroup comparison. The only way that the extinction of the ancestral polymorphism can result in two equally apomorphic characters, or a judgment that "neither resultant state in this circumstance is more apomorphic than the other," as claimed by Wheeler and Platnick, would be a situation where there was a polymorphism between the plesiomorphic character and two different apomorphic characters within the same lineage. Subsequent fixation of one novelty in one lineage and the other novelty in the other lineage would then lead to the conclusion that neither is more apomorphic than the other. This may happen, but it is not the example outlined by Wheeler and Platnick. Thus, only the fact that ancestors may be present and diagnosed should cause Wheeler and Platnick to reject the notion of the apomorphic species concept or the notion that species must be monophyletic. We agree with them on this point: species do not have to be monophyletic because ancestral species cannot be monophyletic.

We submit that the Phylogenetic Species Concept *sensu* Wheeler and Platnick (this volume) is deficient because it does not tie species to speciation in any direct manner. If binominals are supposed to represent species and species are supposed to be the outcome of various processes termed *speciation,* then we would expect some relationship between the application of binominals and the processes of nature. One process that all phylogeneticists recognize is cladogenesis. From what we can deduce, adoption of the Phylogenetic Species Concept *sensu* Wheeler and Platnick (this volume) leads directly to divorcing the application of binominals from any association with cladogenesis. This is because speciation is equated with character fixation and not with lineage branching. Divorcing binominals from cladogenesis was advocated by both Simpson (1961) and Mayr (1982; see comments by Kimbel and Rak 1993). While this practice might be fine for evolutionary taxonomists, it is unwise for phylogeneticists for two basic reasons.

Divorcing binominals from cladogenesis destroys the distinction between tokogeny (nonhierarchical descent) and phylogeny (hierarchical descent). For the Phylogenetic Species Concept, speciation is permitted to occur within a tokogenetic array (a lineage) whenever an apomorphy is fixed. Thus, the application of a binominal does not represent an acknowledgment of the transition between tokogenetic and phylogenetic relationships. It also generates a number of problems on a practical level that get worse as we gain new information. Consider Wheeler and Platnick's example shown in their figure 5.2B. Although we may have no problems sorting out species 1 and species 2 at the top of the tree, what do we call the tokogenetic relatives of species 2 between the cladogenetic event and the final extinction of the last white ball plesiomorph? Is it species 2? It cannot be species 2 because speciation does not occur until the last of the white balls expires. Is it species 1? If so, then cladogenesis is as irrelevant to phylogenetics/cladistics as it is to evolutionary taxonomy. The same reasoning applies to the example in their figure 5.2C.

Divorcing binominals from cladogenesis destroys the basis for the assumption that sister taxa have the same time of origin. The proposition that sister taxa have the same time of origin is fundamental to the phylogenetic system. It is the basis for the assertion that sister taxa should have the same rank in a Linnaean classification, and it is the basis for asserting that there may be some commonality between earth history and clade history. To put it simply, the assertion that sister taxa have the same time of origin is fundamental to the assertion that they are comparable entities. But consider Wheeler and Platnick's figure 5.2C. Species 2 and species 3 may have times of origin that are completely different because their times of origin are not correlated with their common cladogenetic event. Consider their figure 5.2B. The time of origin of species 2 may be thousands of generations after the cladogenetic event. (As an aside, we should point out that species 1 and species 2 in this figure are not sister species—one is the ancestor of the other.)

This practice has unfortunate consequences when we turn to biogeography and historical ecology. One of the fundamentals of vicariance biogeography is that groups can be compared if they live in the same region and if they have responded in similar ways to common events in earth history. Less ponderously, unrelated clades may have common biogeographic patterns because they have vicariated together. Vicariant patterns are cladogenetic patterns. But how are we to compare different clades and seek general pattern and process if there is no necessary correlation between their nomenclature and their cladogenetic histories?

What happens if several different transformation series have characters evolving at different rates? Consider the following scenario. A particular lineage exhibits polymorphisms in seven different and epigenetically uncorrelated transformation series (also known as characters or matrix columns). In each, evolutionary forces are driving the apomorphic character toward fixation, but at different rates. Will we end up subdividing the anagenetically evolving lineage into seven different species, one each time an apomorphy becomes fixed? This might be what Simpson (1961) would do, but does this serve the best needs of phylogenetics *sensu* Hennig (1966)? We think not.

We suspect that these points do not interest Wheeler and Platnick. We suspect that they are really interested in associating binominals with characters. Like all transformed cladists, Wheeler and Platnick build their system around the proposition that characters form hierarchies. The characters and their hierarchies are the important things. The binominals come along for the ride, being names that "express succinctly predictions about character distributions." If characters did form hierarchies, then the transformed cladists would have a strong position. Species as character bearers might be important only in so far as they accurately reflect the natural hierarchy of synapomorphies. The hierarchy of classification would then reflect a hierarchy of character evolution accurately even if some sacrifices must be made in reflecting cladogenesis. Besides, most of the cladogenesis would be represented well enough for most purposes.

We take a different view. We assert that characters cannot and do not form true hierarchies. Thus, we assert that character "hierarchies" do not have the attributes that either Woodger (1937) or Hennig (1966) ascribed to hierarchies. To put it bluntly, characters do not form hierarchies because they cannot. Our reasons require some discussion.

The strong claim made by Wheeler and Platnick is that characters form hierarchies *sensu* Woodger (1937). Woodger (1937) characterized a hierarchy as a system composed of the set of all one–many relationships which have one and only one beginner. Later, Woodger (1952a) termed this definition insufficient and expanded it. He also mentioned the kinds of relationships that exist between elements of a hierarchy. There are three: (1) the "part of" hierarchy, where one thing can stand in hierarchical relationships to its parts (organ to organism, U.S. Congress to federal government); (2) the "father of" hierarchy, where one element can give rise to another element (a generational hierarchy such as an ancestor-descendant relationship); and (3) the converse of the relationship of immediate inclusion of classes (example given, a Linnaean hierarchy). Woodger clearly saw Linnaean taxonomy as a class construct, but he noted that if species are evolutionary entities (concrete particulars) and if genera are also evolutionary entities, then species would be parts of the genera to which they belonged. Thus, the Linnaean hierarchy would be a "part of" hierarchy.

It is notable that Woodger (1952a) did not discuss the characters of entities; his examples are of the entities themselves. This is not surprising because Woodger did not think much of characters (Woodger 1952b, 1953). He considered such statements as "red hair is handed down from parent to offspring" to be metaphorical (see discussion about "Tom's red hair" in Woodger 1953:319–320). Also note that Hennig (1966) did not discuss hierarchies in terms of characters but in terms of entities.

We note, in passing, that Woodger (1961) does provide a discussion of taxa and characters. In this work, he treats taxa as sets polythetically defined by characters. He uses certain logical formalisms to demonstrate how it is possible to justify the gradual evolution of reptiles from amphibians since it is logically possible to proceed polythetically from the "core" of Amphibia to the core of Reptilia, gradually substituting one reptilian character for one amphibian character until the entire suite of reptilian characters has evolved. If taken seriously, this could justify Simpson's (1961) concept of minimum monophyly and for what phylogeneticists could consider the polyphyletic origin of monophyletic groups. However, we do not take it seriously because we do not consider taxa as classes, but as concrete particulars (i.e., individuals).

We believe that the strong claim of Wheeler and Platnick is wrong, both in terms of Woodger's (1937) original characterization and in terms of the manner in which he later characterized hierarchies (Woodger 1952a). Our objections are enumerated below and cause us to conclude that character hierarchies are metaphors.

1. At least some characters may not have their origins as *unique beginners*. We do not know the relative number of synapomorphies that have multiple beginnings due to recurrent mutations in Mendelian populations, nor do we understand the dynamics of how such changes became unique evolutionary innovations. Certainly some have unique beginnings in single individual organisms and others do not. Unless apomorphies act in a manner different from that of other characters, their fixation in lineages is a process involving tokogenetic rather than phylogenetic processes; thus, their origins as apomorphies at the phylogenetic level (homologies diagnosing lineages) are nonhierarchical phenomena.

2. Characters do not give rise to other characters. Thus, characters do not participate in any generational hierarchy for which the relationships are "father of" relationships. Statements in the form of "character A gave rise to character A'" are mere metaphors, not actual statements of ancestry and descent. Why? Because either semiconservative replication or ontogeny comes between one character and its homologous "descendant." This is true of parent and child and thus must be true of ancestral species and their descendants. Ancestry is a characteristic that can be shared between biological entities, not between the parts of these entities as we understand them. Ancestry between

characters is a metaphor for ancestry between entities such as species or monophyletic groups. Of course, it is a useful metaphor because the genetic and/or epigenetic mechanisms that underlie generation of characters provide the rationale for considering these very characters as evidence for corroborating the relationships among taxa.

3. Characters are parts of organisms, but a character of one organism is not part of the homologous character of another organism. Thus, characters cannot participate in a "part of" inclusion hierarchy.

4. Homologous characters (apomorphies; Wiley 1975) are a curious lot, and their nature is actively debated (cf. Patterson 1982; Rieppel 1988; Wagner 1989; Haszprunar 1992; McKitrick 1994). Viewed from the atomic and molecular levels in terms of structural makeup, there is no "homology." The right hands of Ed Wiley and Rick Mayden are composed of molecules that have no direct historical connection (common ancestry) that is relevant to their being homologous. Even the DNA that ultimately arbitrates our epigenetic systems comes from varying sources that are quite unconnected historically (i.e., this nucleotide was synthesized from cow muscle but that one from lettuce leaf). Even the organ systems produced by ontogeny are historically heterogeneous. Viewed from the cellular level, most organs are paraphyletic or even polyphyletic because their organogenesis involves cell lines that are not each other's closest relatives in the "father of" hierarchy of cells. What are connected historically are the cell lineages of an organism linked by mitosis and participating in epigenesis, organisms linked by reproductive ties, and species linked by common ancestors. There is a continuous thread that makes the hierarchy of life, but that thread is a thread of entities participating in processes (ontogeny, tokogeny, and phylogeny), not a thread of characters giving rise to descendant characters. It is the entity thread that allows historically unconnected bits to be assembled into the historically relevant characters that we term *apomorphies*.

We conclude that statements of homology in systematics are interesting only as they relate to specific hypotheses concerning the relationships of entities. Ontogeny is not an independent criterion because character formation is not hierarchical but reticulate. We conclude that certain entities form hierarchies. Because the reconstitution of homologies with every generation is due to semiconservative replication and/or plesiomorphic genetic or epigenetic pathways, there is a correlation between the attributes of organisms and their hierarchical relationships. If this were not true, we would be out of business or we would be doomed to continual frustration. Then perhaps we could vote on the matter. However, the nature of homology is such that characters cannot form real hierarchies; thus, character cladograms are not independent of phylogenetic trees. If our reasoning is sound, then any assertion that cladograms have primacy over phylogenetic trees is not correct. Cladograms must be phylogenetic trees because if they are not, then they are simply metaphors.

Phylogenetic Species *sensu* Mishler and Theriot

Mishler and Theriot continue the argument first presented by Donoghue (1985) and Mishler (1985) that species be treated as if they have the same ontological status as monophyletic groups of species (natural higher taxa *sensu* Hennig 1966). We disagree with this proposition and do not propose to discuss it further except to say that if species are ontologically identical to higher taxa, then both species and higher taxa can be ancestors, given the fact of evolutionary descent. (Of course, if one is a creationist, there are other possibilities.) Evolutionary taxonomists might love this line of reasoning because it seems to provide a justification for some sort of ancestral Reptilia. After all, if only a part of a species is the "real" ancestor, then why not only part of Reptilia as the real ancestor of birds? Using this line of reasoning renders Reptilia monophyletic. This line of reasoning has been rejected by Wiley (1981b) because it produces a logically inconsistent relationship between taxonomy and phylogeny. We submit that Wiley's (1981b) analysis also applies to the arguments of Mishler and Brandon (1987) and Mishler and Theriot. We do not need another logically inconsistent concept of taxa. We do not need artistic judgment in assessing the "significance" or "worthiness" of entities so that their place in the taxonomic hierarchy can be determined.

Are there arguments we like in the paper by Mishler and Theriot? Yes. We agree that reticulation is not a phenomenon unique to species. We agree with much of the discussion concerning asexual and sexual reproduction because these modes of reproduction relate to the nature of species. We do not see, however, how these arguments justify a preference for this version of the phylogenetic species concept.

Biological Species *sensu* Mayr

Probably more attention has been focused on the Biological Species Concept than on any other species concept. Repeating the numerous objections that have been raised in the past, or reviewing the numerous defenses that have been conjured up by advocates, is beyond the scope of this short reply. So we will keep our comments succinct.

Mayr equates speciation with the acquisition of reproductive isolation. He also characterizes the Biological Species Concept as both a relational concept and a nondimensional concept. We discuss each of these qualities in turn.

We acknowledge that some entities christened with binomials have all of the attributes Mayr attributes to his Biological Species Concept. The question is, are these the only sorts of entities that participate in phenomena such as speciation? We think not. Let us consider sexually reproducing species. If speciation is usually an allopatric phenomenon (Lynch 1989; Grady and LeGrande 1992; Chesser and Zink 1994), then speciation does not "require" reproductive isolation to evolve because the newly evolved species live in different (and frequently adjacent) regions. We have discussed several examples of this phenomenon for North American freshwater fishes (Wiley and Mayden 1985). Mayr might argue that such species are not "interesting," but that is an opinion and not a deduction. It amounts to saying that the products we should be most interested in studying (newly evolved species) are not interesting. The situation only gets worse when we consider alternate modes of speciation. Consider other North American freshwater fishes. Speciation via hybridization, common in plants, has been documented to occur in cyprinids (minnows), poeciliids (live-bearing topminnows), and atherinids (silversides) (Echelle and Mosier 1982; Echelle et al. 1983; Meffe and Snelson 1989; De Marais et al. 1992; Dowling and De Marais 1993). Yet, even in the face of a widespread propensity for interbreeding among species in these families, the ancestral species themselves have managed to maintain their identities while yielding new diversity. (To put it in the most pedantic sense, the keys still work.) The situation becomes even worse when we consider asexual and parasexual entities. What do the binominals mean when applied to entities such as *Escherichia coli*? Do binominals apply to such entities? If so, are such entities important from the evolutionary perspective? We know that phylogenetic trees can be prepared for asexual and parasexual entities. Indeed, such entities seem to form monophyletic groups that can only be the historical by-product of ancestor-descendant relationships. The Evolutionary Species Concept seems quite capable of ascribing biological significance to such diversity; the Biological Species Concept seems incapable. Thus, we conclude that the Biological Species Concept, at its best, reflects part and only part of a larger concept, the Evolutionary Species Concept. We also concur with previous authors (cf. Frost and Hillis 1990) that at its worst, the Biological Species Concept seriously underestimates the number of entities that should be recognized through the application of binominals because they appear to be the historical result of speciation. We note, in passing, that Mayr and Ashlock (1991:86–109) do advocate recognizing allopatric but closely related taxa as species, and they recognize asexual species as worthy of binominals (Mayr and Ashlock 1991:32–33). Thus, even those systematists closely associated with the Biological Species Concept seem willing to apply binominals to what appear to be evolutionary rather than biological species.

The fact that the Biological Species Concept is both a relational and a nondimensional concept poses several difficulties. We submit that the ontological status of species as individuals precludes relational species concepts. An entity either exists or it does not. Its existence does not depend on the presence or absence of another entity to provide its reality. The moon exists and would exist even if no other moons had been discovered. Vertebrata exists, and its existence does not depend on the presence or absence of its hypothesized sister group, Cephalochordata. If life had resulted in only a single species, would that species not exist because it is the only living entity? We think it would exist. The Biological Species Concept is inherently relational because it depends on a specified attribute (reproductive isolation) to exist between species, thus requiring at least two species for the comparison. The Hennigian Species Concept goes the Biological Species Concept one better: the relational quality must be between closest relatives. Various forms of the Phylogenetic Species Concept are also relational because they depend either on the presence of an apomorphy or on the attribute of diagnosability. Both attributes render these concepts relational. We acknowledge that many terms are relational: sister group, brother, ancestral species, and so forth. But these are natural classes that describe particular attributes that exist between entities. The entities must exist before they can have a relationship. Furthermore, we acknowledge that relational properties must exist within entities. Strong and weak forces are relational properties that exist within molecules, gravity is a relational property, and sexual reproduction is a relational property among males and females that are parts of many species. Like Paterson (1993), we see no reason to define entities on the basis of relational properties that exist between the entities in question. Entities either exist or they do not. And whether they exist should be determined by their intrinsic qualities, not by some supposed or real relational property between entities.

We find the nondimensional property ascribed to the Biological Species Concept one of the most curious of its properties. If the Biological Species Concept is truly nondimensional, then it is not the concept on which to base the integration of systematics and evolution because it cannot serve as a conceptual basis for the mapping function that is necessary to integrate the tokogenetic and phylogenetic aspects of evolution. A dimensional concept, a lineage concept, necessarily fulfills this function.

Hennigian Species *sensu* Meier and Willmann

We agree with many of the points that Meier and Willmann make. Species should be delimited by speciation events, and the mark of speciation is the transition from tokogenetic to phylogenetic relationships. Lineages cannot be subdivided into arbitrary segments. We also disagree with several key points. These are taken up below.

Much is made by Meier and Willmann about the necessity for ancestral species to always go extinct at the vicariance event (the lineage branch point). From what we can tell, there are two reasons for this necessity. First, because species are relational entities defined in reference to a specific reproductive gap, the ancestral species cannot be the same as either of its descendants because the original reference point (the reproductive gap between the ancestor and its own sister species) no longer exists. We submit that if species are not relational entities, then there is no such logical necessity. Second, persistence of ancestors does violence to the very concept of monophyly. This would be a very serious change, if true. The logic goes something like this. Take four species, an ancestor and three descendants related as shown in figure 11.1a. Meier and Willmann would submit that species X and Y form a monophyletic group by virtue of the fact that they are later descendants of ancestor 1 than is species Z. Because the

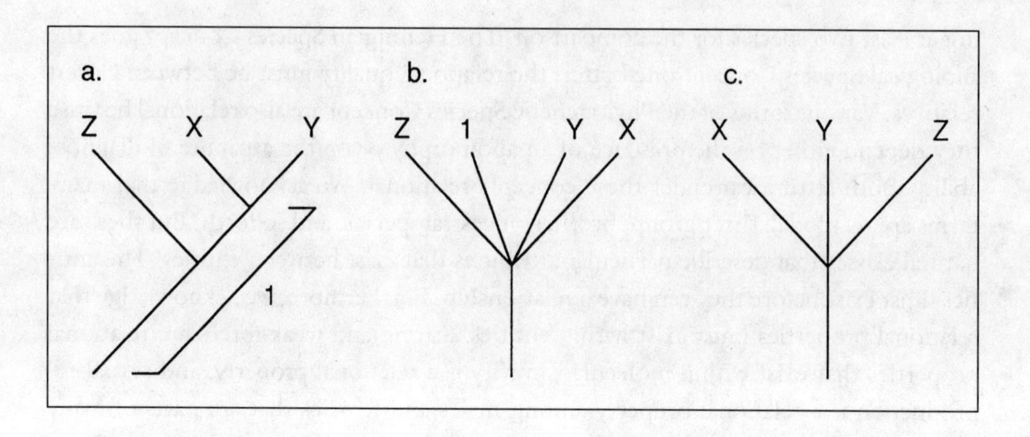

FIGURE 11.1.
A phylogenetic tree (a) and two trees that might result from an empirical analysis (b, c). The phylogenetic tree is considered true in terms of relative time of branching but carries connotations about character evolution (i.e., although the times of origin of species X, Y, and Z are mapped as dichotomous cladogenetic, this may or may not have been accompanied by character change). Trees b and c are two possible trees supported by empirical character data given certain assumptions discussed in the text.

survival of species 1 prevents species X and Y from having a unique ancestor, and because this is required by the definition of monophyly, species X and Y cannot form a monophyletic group. Thus, goes the reasoning, ancestor 1 cannot survive the first branching event if the monophyly of species X + Y is to be preserved.

We suggest that there is a flaw in this argument. It is the assertion that species X and Y necessarily form a monophyletic group exclusive of species Z. This is analogous to asserting that son 2 and son 3 are more closely related to each other than either is to son 1 just because they appeared later in time. In fact, all three sons are equally related to their father. In theoretical terms, if ancestor 1 survived the first speciation event and then gave rise to species X and Y, then species X and Y do not form a monophyletic group because they are an incomplete sister group system. Only the group of four species (1, Z, X, and Y) would form the monophyletic group because only this taxon fulfills the condition of monophyly: an ancestor and all descendants of that ancestor. In practical terms, we would expect an analysis of character evolution of such a scenario to be an unresolved quadchotomy (figure 11.1b, ancestor and descendants in the analysis) or an unresolved trichotomy (figure 11.1c, only descendants analyzed). If character evolution took place between the first and second branch points, then the polytomy would be resolved and it would be assumed that the ancestor (1) went extinct at the branching point. Would this assumption be biologically correct? Probably, but it would take a time machine to find out.

We view the controversy over the persistence of ancestral species as an open philosophical question. It is possible that conceiving of species as lineages and individuals must logically lead to the necessary extinction of ancestral species each time there is a cladogenetic event. So be it. But we have not heard that argument made in any convincing fashion. Because we view species as individuals, we think that there are compelling reasons to think that particular species are defined nonrelationally. Thus, the argument that the relational nature of species leads logically to extinction of ancestors carries no weight. Because we know that other logical individuals can survive their descendants and that doing so does not pose any particular logical problems for their relationships with one another (sons 2 and 3 in the argument above), we do not see the logical necessity in terms of individuality or monophyly. In practical terms, in the world of the real application of binominals, we do not think there will be any nomenclatural changes even if everyone agreed with Meier and Willmann. Hennig conceded that point (see the discussion concerning *Stenodiplosis geniculati*; Hennig 1966:61).

Finally, we do know when species cannot be postulated as surviving ancestors. Wiley (1981a:98–109) outlined the various kinds of phylogenetic trees that can result from mixing different levels of biological organization (populations, species, monophyletic groups, etc.). On the empirical level, specimens taken to represent ancestral species must have the synapomorphy or synapomorphies that diagnose their descendants. No synapomorphy, no hypothesis of ancestry. In addition, an ancestral species cannot have any autapomorphies. So, there are many situations where the empirical evidence leads to a rejection of a hypothesis of ancestry. First, monophyletic groups cannot be ancestral because if they were, then they would be paraphyletic (Wiley 1981a, 1981b). Second, groups of species that are analyzed and found to have purely dichotomous relationships may contain ancestors, but we have no way of proving this with any empirical data, so no ancestors can survive in these situations. In fact, the only situation where the empirical evidence leads one to think that persistent ancestors might have survived a speciation event are cases involving polytomies with species that lack autapomorphies altogether. Much evolutionary investigation would have to be undertaken to verify or falsify cases involving polytomies as possible cases of ancestral persistence. In practice, dichotomies will cause investigators to assume ancestral extinction, while persistent polytomies will lead investigators either to assume that a lack of resolution is due to their inability to find the "right" characters or to assume that they have encountered a persistent ancestor. Which of these assumptions is the reasonable one would be open to question on a case-by-case basis and might well involve data apart from character data.

Our concluding remarks concern reproductive isolation. If the criterion of absolute reproductive isolation between sister species is the basis for Hennigian species, then we are going to have to get used to a vastly fewer number of species than we presently recognize. As ichthyologists, we know of no recently evolved and closely related species of North American freshwater fish that is 100% reproductively isolated from its sister species. What do we mean by closely related? We mean species pairs that are young enough and whose biogeographic relationships are such that there is no reason to think that there is an extinct sister species left out of the analysis (see Wiley and Mayden 1985). Many such species live in parapatry or allopatry, just as predicted by Wallace (1855). We can produce F^1 hybrids if we have a mind to do so. Yet, they are hierarchically related, and their cladogenetic histories are correlated with specific and understandable events in earth history. They act as hierarchical entities, not tokogenetic entities. We submit that they are species if for no other reason than the fact that their histories seem to portray a picture of phylogenesis rather than tokogenesis.

PART 3

Reply Papers (Rebuttal)

12

A Defense of the Biological Species Concept

Ernst Mayr

Every naturalist and every evolutionist is awed by the diversity of living nature. One cannot help asking oneself why there are so many species. Indeed, why are there species at all? Why is not the organic world a single continuity? Why has nature, and more precisely natural selection, favored the discontinuities among the species? What is the meaning of species? The answer the evolutionist gives to these questions is the Biological Species Concept.

Reading through the proposals of various phylogenetic, evolutionary, and other species concepts, I am disappointed that these putative concepts do not give answers to these questions. Rather, these so-called concepts are merely operational prescriptions for the demarcation of species taxa. Neither the phylogenetic, nor the evolutionary, nor any of the other species definitions describe the role that species play in the processes of the living world, and it is on this that a species concept must be based. In fact, most of these essays fail to provide a clear discrimination between a species concept and a species taxon. I refer to Bock (1995) for a perceptive analysis of the meaning of these two terms. It is therefore not surprising that most of the criticisms of the Biological Species Concept seem to be caused by a disappointment of the critic that the Biological Species Concept is not a set of instructions for the demarcation of species taxa. It will help the reader understand my ensuing comments if this important point is kept in mind.

Uniparental Organisms

Mishler and Theriot criticize that the Biological Species Concept does not account for uniparentally (asexually) reproducing organisms. The Biological Species Concept obviously does not fit such organisms. It is futile to argue about asexual species (agamospecies) until those who work with such species develop an appropriate species concept. Agamospecies do not form populations, as defined by the biologist. Because they usually are not subject to genetic recombination, the gene pools of these clones do not require protection by isolating mechanisms. Admittedly, such asexual organisms exist in nature, and there is a general desire among the specialists in such groups to be able to rank them in the Linnaean hierarchy. However, it is not my task to define the agamospecies, nor am I qualified to do so. The only thing that is obvious is that asexual organisms cannot be accommodated by the Biological Species Concept. Up until now, we have instructions on how to delimit asexual species taxa but no concept.

What kind of work should a taxonomist do? For instance, should he drop everything else in order to name previously undescribed species? I have discussed this question in Mayr and Ashlock (1991:14–18). The answer is that, on the whole, new species should be described only in connection with generic or species-group revisions. Exceptions would occur if a species must be named in connection with conservation work or if a modern revision is available. However, doing revisions of previously poorly studied groups would seem more meritorious than making cladograms on the basis of the existing literature.

Allopatric Populations

No matter what species concept is adopted, the species status of allopatric (and allochronic) populations can be determined only by inference or by subjective criteria. This is not a special weakness of the Biological Species Concept. Wheeler has written that other species concepts, such as the Phylogenetic Species Concept, do offer criteria by which to recognize allopatric and allochronic populations, but to recognize such populations is indeed easy. What is hard to determine is whether or not they have reached species status. All the operational criteria that I have found in the literature by which to make a demarcation of such species taxa were subjective.

Reproductive isolation of a geographically isolated population cannot be observed, of course. Indeed, prior to the first third of the twentieth century, the majority of authors simply raised each such population to species rank. The result was chaos. The total biological inequality of such species eventually led to the rejection of this option and to the wide acceptance of polytypic species taxa.

My assertion that, for most local biological situations, it is irrelevant whether a far distant allopatric population is ranked as a species or a subspecies does not mean one should not attempt to rank them on the basis of the best available evidence. Indeed, I have done so conscientiously in all my taxonomic revisions. However, to repeat, the ultimate decision is irrelevant in local biological studies.

Subspecies and Typological Species Concepts

I rejoice in the contention of Wheeler and Platnick that "well-defined, diagnosable subspecies should simply be called species" because it confirms that they have a strictly typological species concept, as I had long suspected from their writings. To explain why vertebrate (and some other) taxonomists recognize subspecies would be superfluous because it is explained so clearly in some standard textbooks (e.g., Mayr and Ashlock 1991:43–45). I realize that it is apparently distasteful for a cladist to read anything not written by another cladist.

At the beginning of the twentieth century, the British ornithologist B. Sharpe, on the basis of a species concept that was not very different from the Phylogenetic Species Concept *sensu* Wheeler and Platnick, recognized about 28,000 species of birds. Any isolated allopatric population (in fact even some nonisolated ones) was called a species. He treated all incipient species and geographic races the same as genuine good species. Later, when polytypic species taxa were recognized, based on the Biological Species Concept, it was possible to reduce the number of real species to about 10,000. Should we now double the number of species of birds on the basis of criteria that basically have nothing to do with the biological meaning of species but go back to the old typological species concept? A return to a typological species concept would probably not greatly affect groups to which the Biological Species Concept had never been applied.

What is now called the Typological Species Concept was usually referred to as the Morphological Species Concept in the older literature. It is to be applauded that this term is now disappearing, primarily because we now have much additional evidence (e.g., molecular, behavioral, sterility) helpful in the demarcation of species taxa and second because the making of inferences based on several species concepts (including the Biological Species Concept) is sometimes helped by using morphological criteria without making such a concept a morphological concept.

The Biological Species Concept has never insisted that the demarcation of species taxa must be based on "observed" interbreeding; the potential for the interbreeding of isolated allopatric populations must always be inferred.

Species as Units of Evolution

I must repeat that the Phylogenetic Species Concepts are not biological (answering the Darwinian question "Why are there species?") but instead only instructions as to how to delimit species taxa. Darwin's "why" question can be answered, like all attempts to give a scientific explanation, only by testing conjectures. Species are the product of evolutionary history, but only the Biological Species Concept is based on an explanatory scenario. Simply to say that species are different is not a concept but refers to the discrimination of species taxa.

In their critique of my conclusion "the species is a device for the protection of harmonious, well-integrated genotypes," Wheeler and Platnick confound the evolution (by selection) of the biological species (its concept) with the origin of particular species taxa.

I have recently discussed in detail what the objects of selection are: gametes, individuals, and social groups (Mayr 1997). Genes are not an object of selection, and neither are species as such. Most natural selection is what geneticists refer to as stabilizing or normalizing selection. All individuals deviating too far from the optimal phenotype are vulnerable and apt to be eliminated. But we do not really understand why some populations seem to be completely uniform, while those of some other species are highly variable. In some genera we have numerous sibling species; in others every species seems to be quite distinct. But because morphological difference is not the species criterion, such differences in variability are irrelevant as far as the species concept is concerned. To be sure, they may cause considerable difficulties in the discrimination of species taxa.

If a population is highly successful, despite showing considerable variability, it indicates that the variation is not disharmonious. If it were, any disharmonious combinations would be inferior and eliminated by selection. It does not follow that a gene pool is disharmonious because it shows much variability.

Species in Time

Willmann and Meier do not seem to understand that one needs a yardstick for a demarcation of species taxa, particularly for the assignment of isolated populations. It is the function of the species concept to serve as such a yardstick. This is as true for the time dimension as for the geographical one. Both Simpson and Hennig tried to find a better method for the delimitation of species taxa in the time dimension but both failed completely. There is nothing in Simpson's evolutionary species definition that permits a demarcation (at the beginning and the end) of species in a phyletic lineage.

Most new species originate by budding, and the establishment of a new founder population is usually the beginning of a new incipient species. Splitting is less common, and because most species originate by budding (even in sympatric speciation), Hennig's suggestion to call the part of a phyletic lineage above a budding point a different species from the part below the budding point is a purely clerical ("bookkeeping") device and biological nonsense. The biological properties of the lineage above the budding point are the same as those below it. I have illustrated this for insular species derived from the New Guinea mainland population of the kingfisher *Tanysiptera* (Mayr 1963:503). I can only repeat that both Simpson and Hennig completely failed to find a solution for the temporal demarcation of species taxa in a phyletic lineage.

The application of the Biological Species Concept to fossils is much facilitated by their usual geographical restriction. As I have often explained, because the Biological Species Concept is strictly valid only in the nondimensional situation, the chopping up of a continuous phyletic lineage into species must be done by inference, as explained in Mayr and Ashlock (1991:106). If there is unbroken continuity, an arbitrary cut has to be made using the same criteria as for geographical isolates (Mayr and Ashlock 1991: 104–109). I do not know of any other species concept that has a superior solution for the division of a long, continuous phyletic lineage into species taxa.

Isolation Mechanisms and Cohesion

Willmann and Meier argue, as had Paterson in his recognition concept, as if the acquisition of isolating mechanisms were a teleological process. Like all other aspects of adaptedness, it is instead simply a by-product of divergence during isolation. There is no cosmic teleology, and therefore nothing evolves "in order to . . ."; yet when perfected, the isolating mechanisms serve to protect harmonious gene pools.

The claim by Willmann and Meier that the existence of reproductive cohesion would permit treating as full species "many subspecific taxa, including populations and demes," is clearly in conflict with the definition of the Biological Species Concept. I do not use the concept of reproductive cohesion, but all populations of a species are capable of interbreeding.

Similarly erroneous is Wiley and Mayden's charge that the Biological Species Concept is misleading because "newly evolved species live in different (and frequently adjacent) regions" and "speciation does not 'require' reproductive isolation." But they fail to demonstrate that these populations are species and to specify the criteria used.

Wiley and Mayden also imply that the Biological Species Concept cannot be nondimensional. They seem to overlook that the noninterbreeding with coexisting species is precisely such a relational aspect. They ask whether a species existed if it was the only one in the universe. Of course, but that is not the problem. What is, is indicated by the question of whether it is legitimate to refer to a unique single entity as a species. What is involved is not its existence but its rank.

Nondimensional means "at the same place and at the same time." The species status of a given population with respect to a different population can be tested only where they coexist in space and time, that is, in the nondimensional situation.

Molecular Evidence

Contrary to claims by Willmann and Meier, nothing would be further from my mind than to suggest that all the describing of new species should now be done by using molecular methods. In the case of suspected sibling species and in the determination of the closeness of relationship of congeneric species the molecular methods are sometimes very helpful. However, it is usually advisable to use several different methods (e.g., both mitochondrial and nuclear genes) to prevent being tricked by mosaic evolution.

As desirable as it would be, we are still a long way from having a unified species concept. A strictly typological species concept is still prevalent in the taxonomy of many higher taxa. Also, there are some real differences among the species of different taxa. The asexual "species" are simply not the same as the biological species of mammals, birds, or butterflies. In most groups of the lower invertebrates, what kind of species they are has never been carefully studied.

Mammals and birds seem to be rather sensitive to hybridization, and one does not find many hybrids in nature. But even here introgression takes place, as for instance from coyote into wolf. Some groups of plants are far more tolerant of introgression than mammals, and may continue to hybridize for millions of years without losing their identity. The tolerance of these plant groups must presumably be explained in terms of their developmental systems. In biology everything is far more variable than in the physical sciences, so the observation that there are certain differences between animals and plants, or between certain taxa of animals, is not unexpected.

Let me repeat, since this is so often misunderstood, that species taxa are multidimensional, but the nondimensional situation is required to determine the crucial biological properties of the species concept.

13

A Defense of the Hennigian Species Concept

Rudolf Meier and Rainer Willmann

"The species concept is crucial to the study of biodiversity. It is the grail of systematic biology. Not to have a natural unit such as the species would be to abandon a large part of biology into free fall, all the way from the ecosystem down to the organism. It would be to concede the idea of amorphous variation and arbitrary limits for . . . intuitively obvious entities . . ." (Wilson 1992:38).

The Hennigian Species Concept describes species as natural entities, that is, as entities that are individual-like products of evolution. The question is, which species criterion is of particular importance in an evolutionary context? We believe that absolute reproductive isolation is this criterion. Its occurrence is a point of no return in the evolutionary process and keeps natural groups of populations apart. The Hennigian Species Concept is rooted in an over 300-year-old biological tradition of applying the term *species* to reproductively isolated units. Below the level of reproductively isolated units, phylogenetic analyses should not be performed because of possible netlike relationships. Reproductively isolated units, on the other hand, are appropriate taxa for phylogenetic analyses because they are hierarchically related.

Natural entities do not evolve in order to be easily recognized. It is therefore to be expected that many species are difficult to identify. It is often difficult to delimit them in space and time and in relation to their closest allies. But it is not our task to present a practical solution to the species question. Instead, in science the only acceptable species concept is one that describes objectively real entities.

Reply to Mayr

Mayr's main problem with the Hennigian Species Concept is the dissolution of the stem species during speciation. He disagrees with our precisely defined temporal species boundaries and criticizes them as arbitrary. He argues (1) that speciation requires genetic change of the populations involved and (2) that in order for the stem species to dissolve, the parental populations have to be divided into two populations of equal size (dichopatric speciation). Before refuting his claims, let us once more stress that the belief that species are real biological units implies that they have a point of origin and a point in time when they cease to exist. The point of origin will be necessarily between or even within generations of the same population. The only alternative is to adopt the logically indefensible position that species are real but have no boundaries in time.

According to Mayr, in peripatric speciation "the phyletic lineage representing the parental species" is not affected by the "budding off" of a new species because "the 'new' species is evidently the same genetically as the old species." We would agree with the general grand-scale scenario, but it is irrelevant as criticism of the Hennigian Species Concept. Two populations or two generations of one population are never genetically the same, because even a difference in a single gene proves them different. Gene frequencies between populations vary continuously, and because it is impossible to draw nonarbitrary boundaries along a continuously distributed variable, any species definition based on genetic similarity is necessarily arbitrary.

However, our main objections to Mayr's point of view are of a different nature. Our species concept and, for that matter, Mayr's biological species definition are explicitly not character based. When Mayr demands that the delimitation of species be based on genetic similarity, he is not only criticizing the Hennigian Species Concept, but his own Biological Species Concept as well. If genetic differences are not part of the species definition, how could the amount of genetic change be of any significance in deciding whether a parental lineage constitutes the same species before and after a splitting event? In both the Hennigian and the Biological Species Concepts, characters are only used as evidence in species recognition and delimitation but are not part of the species definition.

To draw a parallel case, a homologous structure is defined in reference to a single origin in evolutionary history and is not based on similarity. Similarity is used only to establish homology among structures and is similarly used as a tool to identify species boundaries. Eliminating characters not only from species definitions but also from theoretical discussions relating to species constitutes the last step toward a truly adequate Biological Species Concept that describes natural phylogenetic units. These units are complete phylogenetic lineages delimited by objective boundaries, i.e., splitting events. Morphological discussions have their place only in discussions of specific taxonomic problems.

Mayr's argument that stem species can survive speciation events involving peripheral isolates while the old species disappears during dichopatric speciation is also flawed. There is no objective way of distinguishing between the budding off of small peripheral isolates and the splitting of an old species into two new ones of approximately equal size. Which percentage would be the boundary between a peripheral isolate and dichopatric speciation? It is very unlikely that populations ever split into two populations of exactly the same size, and it is even more unlikely that both are dramatically different genetically from the ancestral population, as is required by Mayr in order for the two populations to be new species. If the size of populations originating in a splitting event were relevant, one would be forced to arbitrarily decide when a daughter population is large enough to constitute a new species.

Let us once more emphasize that the distinction between splittings *sensu* Mayr and budding off of peripheral isolates is beside the point, because species is a relational concept, and species are defined relative to their sister species. The number of individuals within either sister species does not affect their relationship. Hence, the size of the daughter species is not mentioned in the species definition of the Biological or of the Hennigian Species Concept. To us it remains, therefore, unclear why Mayr considers this issue important for deciding on the temporal boundaries of species.

Mayr's attacks on the temporal boundaries of Hennigian species stem from his confusion about the applicability of reproductive isolation through time. Because reproductive gaps can isolate contemporaneous populations only relative to each other, they necessarily structure biological diversity only in one time plane. Naturally, reproductive isolation never acts between two subsequent generations. Any application of the concept of reproductive isolation to the time axis implies the extinction of the ancestral, and subsequent reappearance of the descendant species. Curiously, Mayr nevertheless considers the lack of reproductive isolation between ancestor and descendants an argument against the Hennigian Species Concept. We have to conclude that Mayr's failure to make the distinction between parental species dissolving during speciation and daughter species is a remnant of typological thinking.

Typological thinking also has invaded his position on phyletic speciation, the transformation of one species into another through morphological change without lineage splitting. There is no such thing as phyletic speciation. It is an artifact reflecting the traditional way paleontologists classify fossils. Many paleontologists continue to accept the notion of phyletic speciation and deny that species lineages can change considerably over time (Willmann 1985a, 1985b). Mayr nevertheless seems to embrace phyletic speciation when he states that "it has been impossible so far to discover any criterion by which a phyletic species can be demarcated against ancestral and descendant 'species.'" No such demarcation exists. Curiously, Mayr later seems to argue against phyletic speciation when he mentions that "on the whole, whenever a biologist speaks of species, he has in mind the product of the process of multiplication of species, not the product of phyletic evolution."

Mayr's critique paper once more fails to clarify his stand on various important issues. He argues that he has clearly stated how species originate, namely "by the establishment of incipient species and the completion of the speciation process." In other words, Mayr states that speciation is over when the speciation process has been completed and leaves open what the establishment of incipient species curtails, when speciation is completed, and indeed what speciation is. We have to conclude that Mayr's Biological Species Concept remains unable to delimit species along the time axis. How could species *sensu* Mayr then be real historical units? How could they be used in a historical discipline such as phylogenetic systematics?

On occasion, Mayr contorts our arguments. We had pointed out that only reproductive isolation can be used as a species criterion, whereas gene flow and cohesion are also found among infraspecific taxa. He complains that Willmann and Meier "cite my emphasis on gene flow in my 1963 book. They fail to mention that in 1970 and in all my subsequent publications I more or less refuted this early viewpoint." However, in the sentence following his own species definition presented in his position paper in this volume, he states, "Alternatively, one can say that a biological species is a reproductively cohesive assemblage of populations." Again, and contrary to his claims, he stresses cohesion over reproductive isolation. It is not sufficient that Mayr "more or less refuted" his earlier point of view. It is necessary that he finally take a clear stand on the importance of reproductive isolation for species concepts and the insignificance of reproductive cohesion in this context.

Reply to Wheeler and Platnick

Wheeler and Platnick's response to our position paper is seriously flawed for three reasons:

1. Reproductive isolation is repeatedly criticized as the species criterion of the Hennigian Species Concept. At the same time, their Phylogenetic Species Concept is similarly dependent on reproductive isolation when it proposes that only "aggregations of (sexual) populations" should be considered as species. After all, these "aggregations" have to be isolated from other such aggregations.
2. Wheeler and Platnick frequently inappropriately use terms that we had precisely defined in our first paper. The result is unnecessary confusion.
3. Wheeler and Platnick fail to distinguish between theory and practice in systematics.

REPRODUCTIVE ISOLATION

Wheeler and Platnick have fundamental objections to reproductive isolation as a species criterion. They argue that "a concept of species will be required that can be used in practice without experimental observation of interbreeding." First of all, even if reproductive isolation may be difficult to observe or infer, it is isolation that keeps natural taxa apart and creates groups of organisms with an individual-like nature. This simple fact of nature cannot be ignored, and practical difficulties in finding reproductively isolated units are irrelevant in discussions about species definitions. Secondly, they overlook the fact that reproductive isolation is not only important for the Hennigian Species Concept but implicit in their own concept. For biparental organisms, they argue that only "aggregations of (sexual) populations" can be phylogenetic species. They even admit that their concept attempts, "like others before it, to recognize the kinds of life that perpetuate more of like kind." In order to determine the limits of such aggregations, "observation of interbreeding" by direct or indirect means is required. Wheeler and Platnick may believe that they are arguing against the Hennigian Species Concept when they oppose reproductive isolation as a species criterion, but effectively they are criticizing their own concept as well. Contrary to their claims, their own Phylogenetic Species Concept is therefore no more operational than the Hennigian concept.

Wheeler and Platnick argue that the Hennigian Species Concept was designed to circumvent problems of the Biological Species Concept by delimiting species in time. But these changes are not designed to circumvent any problems with the Biological Species Concept, nor are they merely methodological. Instead, the proponents of the Hennigian Species Concept recognize the importance of reproductive isolation for delimiting species living in one time plane. It naturally follows that the origins of reproductive gaps determine the origins of new species. Therefore, stem species must dissolve during speciation. These are simple logical conclusions that follow from recognizing reproductive isolation as the force behind structuring biological diversity. Wheeler and Platnick fail to understand the critical importance of reproductive isolation when they consider this dissolution "a methodological convention."

FLAWED ARGUMENTS

Wheeler and Platnick frequently argue against species concepts from the point of view of a proponent of their version of the Phylogenetic Species Concept instead of evaluating the internal logic of the competing concept as a whole. For example, they state that "because their [Meier and Willmann's] ancestral and daughter species can be identical in all respects, there is no justification for conceiving of them as different species." We provided a detailed discussion about the relational nature of species, and Wheeler and Platnick should have pointed out logical inconsistencies in our concept instead of complaining that our concept is not identical to theirs. Furthermore, what Wheeler and Platnick consider a problem of the Hennigian Species Concept similarly applies to their Phylogenetic Species Concept. According to their Phylogenetic Species Concept, even one and the same individual can belong to two different species. All individuals of an ancestral phylogenetic species 1 that survive the moment of fixation of a new character combination that creates descendant species 2 suddenly belong to the descendant species 2. They were born as members of species 1 and will die as members of species 2 (see figure 5.3 in Wheeler and Platnick's position paper).

Wheeler and Platnick claim that we "define Hennigian species, after all, on the basis of presumed dissolution of the ancestor, not upon properties of sister species." However, sister species are recognized based on characters (synapomorphies). Whenever a new species comes into existence through a splitting event, it is implied that the ancestral species dissolved, because a new reproductive gap had evolved. This is not a "belief divorced from evidence," as claimed by Wheeler and Platnick, unless they want to argue that one cannot recognize aggregations of (sexual) population pairs and reproductive isolation.

Wheeler and Platnick frequently criticize the Hennigian concept for flaws that do not apply to our species concept and should be directed against some misguided ideas about the Biological Species Concept instead. For example, Wheeler and Platnick fail to recognize the difference between isolation and separation, although the important differences have been addressed at length in numerous publications issued within the past 40 years: "Isolating mechanisms are biological properties of individuals which prevent the interbreeding of populations that are actually or potentially sympatric. This definition clearly excludes geographic isolation. . . . Isolating mechanisms always have a partially genetic basis" (Mayr 1963:91). Isolation mechanisms are intrinsic to the organisms involved, whereas geophysical obstacles are not isolation mechanisms but cause the mere (geographic) separation. Therefore, our discussion of isolation mechanisms never refers to separation. To borrow an example from Wheeler and Platnick, just because a few deer are kept in the basement of the American Museum, they are not reproductively isolated but merely separated. They are not a new species.

Ignoring the different meanings of the same terms as used by others, Wheeler and Platnick repeatedly use their own definitions of terms to judge the merits of other concepts. For example, they discuss our objections to morphological characters as defining attributes of species and assume that we are using Wheeler and Platnick's character concept. Their concept may be restricted to features that are constantly distributed throughout populations, and therefore characters *sensu* Wheeler and Platnick may not dissolve throughout time. However, characters in a traditional sense certainly do. Here we need only point to our discussion of quantitative characters in our critique paper. Quantitative characters can be used to distinguish species whenever two populations differ significantly with regard to the size of a structure. Yet, Wheeler and Platnick's character concept is unable to deal with quantitative characters over time because quantitative characters usually undergo continuous evolution, and there is no point along the time axis where the differences between subsequent populations are discrete.

Wheeler and Platnick frequently criticize some version of a Hennigian Species Concept that is distinctly unlike the one that we proposed in our first essay. For example, they claim that we "insist on the separate identities of mother and daughter species that were never isolated." However, we clearly stated in our position paper and again in this paper that the isolation criterion is nonapplicable along the temporal sequence of generations (see also the extensive discussion in Willmann 1985a). Similarly, Wheeler and Platnick pretend that deer separated from the main populations in the basement of the American Museum are isolated, although it is common knowledge that isolation mechanisms are intrinsic to the organisms and separation is not (see discussion above). This incorrect use of *isolation* and *separation* leads to nonsensical statements such as "Meier and Willmann are necessarily vague about when a reproductive gap becomes a species gap. Some gaps open, only to subsequently close in the face of renewed gene flow." Gaps that open and close ad libitum are induced by separation, and separation does not necessarily result in reproductive isolation. The Hennigian Species Concept is not vague about what constitutes reproductive isolation. It is the absolute lack of any gene flow.

Wheeler and Platnick also argue that "intermittent, low-level gene flow among partially isolated populations renders decisions regarding the status of biological species necessarily subjective." This objection ignores our explicit statements that only absolute reproductive isolation can be used to delimit species in one time plane ("It cannot be overemphasized that absolute isolation is the only species criterion that excludes any arbitrariness"). Any gene flow between populations will render them conspecific. There is no subjective element to the decision. However, their misguided discussion about separation, reproductive isolation, and so forth makes us wonder once more how the "aggregations of (sexual) populations" in Wheeler and Platnick's Phylogenetic Species Concept are defined. Given that the deer in the basement of the American Museum are a family group, they probably have some unique genetic markers and would probably constitute phylogenetic species *sensu* Wheeler and Platnick.

THEORY AND PRACTICE

Many of Wheeler and Platnick's objections reflect an approach to science that is entirely different from ours. From our point of view, concepts and their definitions create the theoretical framework for empirical work. A distinction between theory and practice seems lacking in Wheeler and Platnick's world; hence, theory and empirical evidence are confused. For example, they argue that because all existing species of a particular taxon are known rarely, one can never be sure whether "true" pairs of sister species have been identified. This problem only relates to practice and does not affect our species concept. Besides, as more evidence becomes available, we approach a correct reconstruction of the phylogenetic relationships and thus the stem species.

In another instance of confusion of theory and practice, Wheeler and Platnick argue that "because the number of missing species between those species included in an analysis varies widely, Meier and Willmann's insistence on comparisons with sister species must be relaxed in virtually every case." Although a species is defined relative to its sister species (theory), it can also be identified considering more distantly related species (practice). After all, the species will also be reproductively isolated to more distantly related species.

Contrary to Wheeler and Platnick's claims, it is not necessary to conduct a cladistic analysis in order to recognize Hennigian species. Species that exist at one slice of time are identified based on reproductive isolation, which can be inferred from characters.

Therefore, the recognition of contemporaneous species does not require knowledge about character polarity. The subsequent cladistic analysis determines the sister-group relationships among the species, thus providing hypotheses about stem species and their *Baupläne*. Fossils can be included in the cladistic analysis. If they do not deviate from the hypothesized character combination of a stem species, they could potentially be part of the stem lineage.

Taking a quote out of context, Wheeler and Platnick claim that we implied "that there is something wrong with recognizing species on the basis of plesiomorphies." We indeed maintain that delimiting uniparental "species" based on plesiomorphies is inappropriate because it obscures the hierarchical relationships within such agamotaxa. However, we never referred to character polarity in our discussion of biparental species. It should be kept in mind that stem species never carry autapomorphies, but they are nevertheless species according to our concept. For our point of view on species recognition, the extensive discussion in our position paper may be consulted.

Of general importance is Wheeler and Platnick's misconceptions about cladograms. They complain that we wish "to take literally the lines that systematists draw connecting species in a cladogram." Given that cladograms are hypotheses about phylogenetic relationships, there are two possible interpretations of the lines. They may stand for hypothetical species evolving through time. In this case, the lines below the nodes connecting sister species do indeed symbolize stem species. The second interpretation is that the lines are entirely fictitious, implying that the aggregations of (sexual) population pairs were spontaneously generated and have no history of descent. A third and nonphylogenetic point of view would hold that the cladogram only summarizes character information and has no evolutionary implications. As phylogenetic systematists, we clearly support the first interpretation.

Reply to Wiley and Mayden

Wiley and Mayden remain uncommitted with respect to surviving stem species. On the one hand, we were delighted when Wiley and Mayden initially conceded that species "should be delimited by speciation events." On the other hand, it is obvious from the rest of their paper that they propose that stem species sometimes survive speciation. We strongly disagree with this notion. Evolutionary history produces phylogenetic relationships. If species X and Y originated in the splitting event (see figure 11.1), then they are sister taxa, with species Z being a distant relative. This necessitates that species X and Y have an ancestral lineage in common that is not shared by species Z. Wiley and Mayden's species concept is unable to reflect this evolutionary history because it insists that species Z survives. That conventional phylogenetic analysis might yield difficulties in reconstructing this history is beside the point. It cannot be overemphasized that evolutionary history is independent of phylogenetic techniques. Our *Melanopsis* example illustrates this point (see figure 8.3). It can be deduced from stratigraphy that *M. gorceixi* and *M. inexspectata* are more closely related to the recent form of *M. praemorsa* than *M. vandeveldi*. A traditional phylogenetic analysis would have been unable to resolve the relationships because no synapomorphies were preserved. It is nonsensical to argue that in this instance only one stem species and one monophyletic group can be accepted because Hennig's phylogenetic technique, the search for synapomorphies, failed to resolve the phylogenetic history. If one were to adopt this approach, hierarchical information about the relationships of species would be unnecessarily disregarded and the border between tokogeny and phylogeny obscured.

Wiley and Mayden's example of the three sons of which no group of two can be more closely related to each other than to the third is misleading. Ironically, it is Wiley and Mayden who stressed the important distinction between tokogeny and phylogeny in a their position paper. It is this very difference that they are ignoring in this example. The relationships among three siblings are netlike (tokogenetic), whereas those among three species are hierarchical (phylogenetic).

Wiley and Mayden argue that many recently evolved North American fish could not be recognized as Hennigian species because absolute isolation is required by the Hennigian concept. They argue that hybridization is not surprising in recently differentiated populations and claim that "we can produce F^1 hybrids if we have a mind to do so." This objection to absolute isolation as a species criterion is meaningless. Gene flow between two populations is not reestablished unless the F^1 hybrids are capable and do indeed backcross with parental stock. Besides, F^1 hybrids that are produced under laboratory conditions have no bearing on natural populations, as we discussed in our comments on crossing experiments in our position paper. However, if two North American freshwater "species" hybridize and the F^1 generation is found interbreeding with the parental populations, the conclusion is simply that the two "species" are in fact only a single one. After all, there is no reproductive isolation, and only the future will tell whether the two populations will become reproductively isolated. The taxonomist who insists on considering them as two separate species today makes unscientific predictions about the future course of evolution.

Reply to Mishler and Theriot

Mishler and Theriot introduce their critique paper with some general statements about the competing species concepts (see also table 9.1). We disagree with many of their assessments. In the "Ontology versus Epistemology" section, Mishler and Theriot claim that several species concepts are based on a priori theories about the speciation process and that they are designed to reflect what the authors think species should be. What are the a priori theories required by the Hennigian Species Concept? The only "theory" that we can identify is the existence of reproductive isolation, which has been confirmed in countless observations and experiments. Reproductive isolation naturally delimits and produces units among life forms, and these units are species. Thus, the origin of reproductive isolation is speciation. There is no arbitrary element or a priori theory involved. Instead, reproductive isolation exists in nature (independently of any species concept), and our species concept necessarily follows.

In their "Special Reality of Species" section, Mishler and Theriot incorrectly claim that "unlike the other four contributions, we [Mishler and Theriot] feel that species are exactly as real as higher taxa—no more, no less." Because reproductive isolation is a real phenomenon, Hennigian species are as real as higher taxa, provided the latter are monophyletic. We have never argued otherwise and reject Mishler and Theriot's unjustified claims.

When Mishler and Theriot discuss the definition of monophyly, they continue to propose a change to the definition of monophyly that they consider "subtle but important." From our point of view it is neither. As discussed in our critique paper, their monophyletic units exclude stem species. Therefore, their monophyletic groups are in fact polyphyletic, and calling polyphyletic units monophyletic can hardly be considered a subtle change.

In the "Primacy of Sexual Reproductive Processes" section, Mishler and Theriot claim that we "clearly view reproductive cohesion or isolation as the major evolutionary force holding species together." We do not. Reproductive isolation keeps natural units apart, and we are not concerned with processes that hold species together. Mishler and Theriot are constantly searching for forces that maintain cohesion within a species. We are more concerned with biological forces that keep units apart and produce taxa that are isolated from each other over phylogenetic time. Such forces create the hierarchically related units that can be used as terminals in phylogenetic analyses.

Mishler and Theriot argue that the concept of reproductive isolation is not universally applicable and cannot therefore be used as a species criterion. Obviously reproductive isolation cannot be a species criterion for uniparental species because the Hennigian Species Concept does not apply. When a uniparental taxon is produced via the reduction of biparental reproduction, one new species and an agamotaxon originate. Because reproductive isolation is the sole species criterion for the Hennigian Species Concept, it cannot be used to further distinguish between taxa within agamotaxa. If bisexual reproduction were universal, a single species criterion could be used throughout the organismic world. Unfortunately, that is not the case.

We believe that there is no satisfactory species concept for the organisms that primitively lack meiosis. Widespread exchange of genetic information leads to a bewildering phenotypic pattern. Forcing these phenotypic patterns into "species" and considering them in any sense equivalent to bisexual species can hardly be considered an advantage of a concept that is applicable to all life forms. Such a concept masks differences and creates the illusion that there are comparable units for bacteria and insects. We do not claim that a species concept based on reproductive isolation applies to the entire organismic world. Neither do we believe that, given the important differences in reproduction across all of life, it would be warranted.

Mishler and Theriot mention that sexual reproduction is rare in diatoms and imply that reproductive isolation cannot be used in this case as a species criterion. However rare sexual reproduction may be, these diatoms are still sexually reproducing organisms, and the concept of reproductive isolation applies. The only unambiguous difference is between sexually reproducing and uniparental organisms. Given that the frequency of sexual reproduction varies, it would be impossible to find an objective criterion distinguishing between organisms that fall under the criterion of reproductive isolation or not. This point is discussed in detail in our position paper.

Mishler and Theriot claim that Hennigian species are not testable for allopatric populations or species. However, in such cases we can use the same morphological cues for species recognition that are also used in the sympatric situation. Therefore, Hennigian species can be inferred, and the species hypotheses become testable as soon as the populations or species become parapatric or sympatric. They can also be studied under laboratory conditions, although results may be difficult to interpret. When organisms from two populations can be bred in the laboratory and the crossing experiments are not successful, the result is good evidence for the existence of two species. Other outcomes may be more difficult to interpret, but they will provide additional cues (see our position paper).

The main difference between Mishler and Theriot's and our position on species seems to be that they overemphasize that "nature is messy and breakdowns in isolation (i.e., any mechanism resulting in reticulation such as symbiosis, partial genetic exchange) occur." Genetic exchange through viruses is irrelevant to reproductive isolation and does not alter phylogenetic relationships. It merely affects the relationship between gene trees and the corresponding species trees. Breakdowns of absolute isolation are extremely rare in most bisexual taxa, and in most cases we believe that close scrutiny will reveal evidence that absolute isolation was never present. From our experience with the largest taxon of biparental organisms (Hexapoda), nature is overwhelmingly not messy with respect to reproductively isolated units. "Messy" taxa are rare and require closer study, but we are convinced that the Hennigian Species Concept applies to them as well as long as sexual reproduction does occur. We believe that it is preferable to have a species concept that is difficult to apply to a particularly difficult group than to have a concept that is not able to reflect the complex structure of nature.

Mishler and Theriot point out that not all individuals participate in reproduction, even within solely sexual clades: "Every degree of frequency of sex exists among populations of different species, ranging from absolute asexuality, through rare fertilization events, to panmixia." According to their concept, individuals from such a species that die before reproducing are still parts of the species. By the same token, one could argue that a larva not carrying the apomorphies that are used to define a species *sensu* Mishler and Theriot would not belong to the same species. But this argument is pointless.

Mishler and Theriot here simply ignore that we have emphasized the necessity to include the genetic potential to interbreed in any discussions about reproductive communities. Furthermore, we would like to point out that our species concept is based on reproductive isolation and not on breeding, as is implied in their statement.

We also would like to dispute Mishler and Theriot's claim that we consider Hennigian species end products of evolution. One may argue that species that go extinct without leaving descendants are indeed end products. All other Hennigian species are not end-products. They were either stem species that gave rise to descendants or else they are evolving and may become stem species themselves. Hence, they are or may become a stage in the evolutionary process.

In our opinion the Hennigian Species Concept is currently the best species concept for bisexual organisms. It describes species as reproductively isolated groups of natural populations that originate with the dissolution of the stem species in a speciation event (occurrence of reproductive isolation) and cease to exist either through extinction or speciation. This definition precisely outlines the spatial and temporal boundaries of species in a nonarbitrary manner and thus describes natural units. It is particularly well suited for phylogenetic studies because the requirement of complete reproductive isolation ensures that the relationships between Hennigian species are hierarchical.

The competing species concepts are deficient in one or another respect. The closely related Biological Species Concept as advocated by Mayr fails to provide temporal boundaries of species, and its proponents usually allow some arbitrary amount of gene flow between species. The Evolutionary Species Concept *sensu* Wiley and Mayden fails to provide criteria that allow a nonarbitrary delimitation of species in both the spatial and temporal dimensions. The Phylogenetic Species Concept *sensu* Mishler and Theriot ignores the fundamental difference between tokogenetic, or netlike, and hierarchical, or phylogenetic, relationships. Furthermore, it provides no objective choice of ranking criteria, according to which the clades that "deserve" species status can be identified. Lastly, the Phylogenetic Species Concept *sensu* Wheeler and Platnick confuses ontology and epistemology. Furthermore, its definition is vague about the meaning of fundamental concepts such as aggregations of (sexual) populations, and we have shown that the concept is not nearly as operational as the authors claim.

A Defense of the Phylogenetic Species Concept (*sensu* Mishler and Theriot): Monophyly, Apomorphy, and Phylogenetic Species Concepts

Brent D. Mishler and Edward C. Theriot

We find little of substance in most of the criticisms directed our way in the previous section of the book (although, of course, we enjoyed all the criticisms directed elsewhere!), but a few criticisms do come close to the mark. The former we will address quickly, and then will explore the latter more carefully, as they are instructive about the differences among us and the possible solutions.

Reply to Willmann and Meier

Although we applaud their specific, logical deconstruction of the Phylogenetic Species Concept of Wheeler and Platnick, Willmann and Meier simply criticize our violations of Hennigian orthodoxy, restate their original opinions without further evidence, and express "surprise" several times at our "peculiar" positions. They raise rhetorical questions that they could have answered by reading our paper closely. They ask if we "agree that absolute reproductive isolation (not separation) exists between groups of biparental organisms." Yes, nearly absolute isolation (but never an absolutely 0% chance of reticulation following arguments presented in our position paper) may exist at some extreme levels of divergence, but in many plants that would be at the level currently ranked at about the family level. As aptly pointed out by Wiley and Mayden (in their critique paper), if absolute isolation were used as a ranking criterion, we would have orders of magnitude fewer species. Furthermore, the extremely "lumped" species so recognized would miss all the interesting evolutionary processes causing divergence. All those thousands of currently named orchids with their wildly different pollination systems, divergent genotypes, and endemic distribution patterns would go into one big multicolored, worldwide pile, just because they happen to have retained the plesiomorphic ability to form hybrids under some conditions.

Contrary to Willmann and Meier's remarks, we do not deny the existence of stem species, although we would prefer the term *stem lineages*. We are not as quick as they to think we can see them all over the place, but they must have existed. However, it is not logically correct to say that the stem lineage is a member of a contemporaneous monophyletic group—it was the whole monophyletic group at a past time. As cogently explained by Hennig (1966:72), "in the phylogenetic system it [the stem species] is equivalent to the totality of all the species of the group." We do not want to deny classification to fossil lineages; one of us (E.T.) works with fossils, and we both classify fossil lineages with extant ones. We do insist that conventions be followed (such as those nicely codified by Wiley 1981a) to indicate the perceived status of those fossils. Assuming we knew we had one, we would not put a stem species into a modern classification without noting it as such.

Our revised concept of monophyly particularly upsets Willmann and Meier, but we submit that they are not clear about what upsets them. The two contrasting views of monophyly in cladistics (i.e., the diachronic definition, "an ancestor and all its descendants," versus the synchronic definition, "all and only descendants of a common ancestor") have both had adherents for a long time. The synchronic view has the most distinguished history and a more solid justification, beginning, contrary to Willmann and Meier, with Hennig himself. The usage of monophyly in the work of Hennig is clearly synchronic—for example, his use of Venn diagrams to show relationships and his definition (1966:73): "A monophyletic group is a group of species descended from a single ('stem') species and which includes all species descended from this stem species." Nelson's influential definition of monophyly is similarly synchronic (1971:471–472): "a group including all species, or groups of species, assumed to be descendants of a hypothetical ancestral species." Our departure from Hennig and Nelson is not in adopting a synchronic definition, but rather in removing the reference to species in the definition, a departure that is not as radical as it might sound.

Because, as we discussed in some detail (and not contradicted by Willmann and Meier), classification is a matter of human convenience, superimposed on the hierarchy of real monophyletic groups, Willmann and Meier (and others including Wheeler and Platnick) are wrong to say that species should be named prior to phylogenetic analysis. The classification in phylogenetic systematics is and should be applied after the phylogeny is reconstructed (see the discussion of the temporal phases in phylogenetic systematics in our position paper, again not contradicted by Willmann and Meier). As pointed out by the brilliant thinker that we both respect (Hennig 1966), the basic element of phylogenetic systematics is not the species, but rather the semaphoront: the character bearer, the specimen. Analysis is required to link the semaphoronts into ontogenetic, then tokogenetic, then phylogenetic (monophyletic) groups. Taxa, including species, are never a given, knowable before any analysis starts.

We actually agree with Willmann and Meier's point that "ranking of taxa is nonsensical in phylogenetic systematics," and we will discuss it in more detail below. Note that the implications of this statement will actually undermine their own viewpoint: it shows that attempting to determine a universal ranking criterion (their "absolute reproductive isolation") does not make sense.

Reply to Mayr

There is little we can reply to Mayr that has not been said before by us and others. His arguments are based on authority alone. He mainly resorts to empty name calling and dogmatism; he simplistically labels his opponents as typologists, nonbiologists, and so forth. We particularly resent his characterization of us (presumably) as "armchair taxonomists." We have both spent many days in the field and then back in the laboratory looking at systematics of real mosses and diatoms (neither of us works on lichens, despite what Mayr seems to think, but if one does know anything about the biology of those complex symbiotic organisms, one would not use them as a paradigmatic example of monophyletic species). The plain fact remains (contrary to Mayr and to Meier and Willmann) that neither reproductive compatibility nor reproductive isolation can be used as a universal criterion for grouping organisms into species across the spectrum of world biological diversity; it is neither empirically practical nor theoretically sound (see our position and critique papers).

Reply to Wiley and Mayden

These authors do manage to state very concisely and clearly the difference between our view and their view (that is shared with the other authors in this book): it does indeed come down to whether "species [are] treated as if they have the same ontological status as monophyletic groups of species." Wiley and Mayden just flatly disagree with equal treatment, and do not wish to discuss the issue further (nor did they provide substantive reasons for unequal treatment in their position paper), but we think the issue is quite worthy of further discussion—we hope the readers of this book will not engage in such hasty dismissal.

Reply to Wheeler and Platnick

Wheeler and Platnick are concerned that "making species dependent on cladistic analysis implies that Hennig's tokogeny-phylogeny distinction was not valid and that *monophyly* has a meaning other than Hennig's." Our short answer is "yes" to both points; a long answer has been offered in our position and critique papers. The only necessary point to reiterate here (see above under discussion of Willmann and Meier) is that phylogenetic classification (including species) must follow phylogenetic analysis rather than precede it. Contrary to Wheeler and Platnick's canards, we do not deny the reality of species—we accord species exactly the same reality as all correctly named taxa.

We were amused reading Wheeler and Platnick's opinions on the unreality of demes and populations. If they are right there, it undermines their favored concept, because as pointed out by us and by Willmann and Meier, Wheeler and Platnick's definition of species uses *populations* as a key term.

Contrary to Wheeler and Platnick's statements here and elsewhere, the presence or absence of characters means nothing to the ontological status of lineages or taxa. Chronospeciation, as allowed by Wheeler and Platnick, is definitely the *reductio ad absurdum* of their Phylogenetic Species Concept. They confuse the vital epistemological role of characters with the ontology of monophyletic groups.

The comments by Wheeler and Platnick about asexually reproductive creatures are bizarre; they state that two or more individuals must exist to form a lineage. Leaving aside the great difficulties population biologists have in defining what an individual is in clonal plants (e.g., Harper 1977), Wheeler and Platnick must have forgotten that cells form clear lineages, marked by apomorphic somatic mutations, within individuals. Even accepting for the sake of argument their specification that it takes two organisms to form a lineage, it does not help much to avoid the inflation of taxa that their concept would require in clonal groups; instead of naming every individual, they would have to name every pair. Practically speaking, many organisms that reproduce clonally are capable of reproducing daily (or even more frequently), so the distinction between naming one individual and dozens with a new mutation may simply be a matter of a few hours.

Wheeler and Platnick's perceptions about an element of arbitrariness in ranking at the species level in our Phylogenetic Species Concept are valid, even if their characterization and solution of this "problem" are misguided, and will be taken up in the general discussion below.

Final Points: Breeding Barriers, Ranking, and Biodiversity

All the other authors in this book feel that it is essential to come up with a completely universal species concept. This seems to be motivated by both a scientific preference for simplicity and practical needs to assess biodiversity (i.e., to tell people how many species there are). Our Phylogenetic Species Concept was roundly criticized for its lack of universally applicable ranking criteria at the species level. The two of us hold slightly different views of the ontological nature of the species rank and would agree that the issue of ranking has not been adequately addressed by proponents of our Phylogenetic

Species Concept in the past. In order to do so now, we will outline and defend our viewpoint as presented in more depth in our position and critique papers:

1. We continue to maintain that there is not a unique level at which the forces uniting a lineage give way, causing divergence. We do not just mean epistemological difficulties in discovering such unique cleavage points (difficult as that is); we mean an ontological lack of such points. There is indeed a dynamic tension between different cohesion/integration processes that must be explored, but there are many processes impinging in any given case, and the results of these processes are often gradual clonal divergences rather than abrupt discontinuities. There may in some cases be a unique level at which all cleavage processes cause abrupt discontinuities that correspond, but not in all cases (the point of Mishler and Donoghue 1982). Furthermore, in each case different processes may be paramount. So, although the level at which divergence becomes dominant in a statistical sense averaged across all the relevant processes could be part of a general ranking criterion (see Horvath 1997), we cannot pretend that there is some unique cleavage point, universally comparable across all living things. There will always be fuzzy edges to this boundary at several fractal levels (like trying to define the edge of the Gulf Stream in the Atlantic—one can see it clearly from an airplane, but not well from a boat and even less well while swimming with a magnifying lens).

2. Unlike monophyletic groups, which are real, taxonomic names are human constructs, to be applied for our own purposes. Interestingly, no one refuted our most basic point that what was at stake in this debate was a taxon problem, not a species problem. Most of us in this book agree that the paramount purpose of taxonomy is to reflect and communicate phylogenetic relationships among taxa (although we would appear to differ a little bit among ourselves in how to go about this). Clearly, species names must be regarded as taxonomic names, and these names must be applied after an initial hypothesis of phylogenetic relationships among organisms is available. Taxa at all levels in the hierarchy refer to real things if they are monophyletic.

Although monophyletic species taxa diagnosable by sound apomorphies are the ideal under our Phylogenetic Species Concept, we do allow (despite the misrepresentations of critics) that some taxonomic names at the species level will reflect non-monophyletic groups. Stem species are a special case, as discussed above, and should only be recognized in a classification using a special convention (we like the one given by Wiley 1981a:223: "placed in the hierarchy in parentheses beside the taxon which contains its descendants"). Likewise, metaphyletic species (i.e., those not known to be either monophyletic or paraphyletic; Donoghue 1985; Mishler and Brandon 1987) should be recognized only in a classification using a special convention (we like the one given an asterisk by Donoghue 1985). We would never recognize known paraphyletic species (i.e., those in which some parts share synapomorphies with an outside taxon).

Therefore, although the species taxa recognized epistemologically by Wheeler and Platnick's Phylogenetic Species Concept and our Phylogenetic Species Concept in a given genus may often be exactly the same, our Phylogenetic Species Concept is superior ontologically because it would be explicit about which of those species are monophyletic and which are not.

3. As pointed out by several authors (including Willmann and Meier; see above), ranking is arbitrary from a phylogenetic perspective. As we said in our position paper, the ultimate solution to the "species problem" may be to get rid of that rank in addition to all formal ranks (see De Queiroz and Gauthier 1992) and just use uninomials for monophyletic groups at all levels (for an argument along those lines see Mishler 1999). For the discussion in this paper, we have been assuming that the current Linnaean system of ranked classifications is to remain in place, and thus were compelled to give suggestions for ranking criteria for the species level, but if one is concerned about arbitrariness in ranking, getting rid of ranks is the only satisfactory solution. Our Phylogenetic Species Concept as we have defined and defended it does not introduce arbitrariness to species ranking; rather, it recognizes and tries to be explicit about the arbitrariness that does (and must) exist.

4. All the authors in this book, including us, are concerned about estimation and conservation of biodiversity. The conceptual matters we are arguing here have great implications for public debates over conservation, and thus we should be very careful to think the concepts through. In our chapters, we have presented a number of reasons for believing that all named "species" are not (and cannot be made) equal in a phylogenetic sense (or any other sense for that matter). This is something that simply must be pointed out to ecologists and other "consumers" of species names, who tend to use any named species as an indivisible and comparable quanta. As has been pointed out by a number of pioneering cladistic conservation biologists (Vane-Wright et al. 1991; Faith 1992a, 1992b), conservation priorities can best be established by a consideration of phylogenetic relationships. Our Phylogenetic Species Concept, with its explicit acknowledgment of the need for application of different ranking criteria in different cases, lends itself to a more rational assessment of biodiversity, one lineage at a time, instead of a mindless counting of species names. We owe it to the broad world out there to get clear about what species and other taxa are, phylogenetically speaking.

15

A Defense of the Phylogenetic Species Concept (*sensu* Wheeler and Platnick)

Norman I. Platnick and Quentin D. Wheeler

The phylogenetic species concept, as we have presented it, represents a logical next step in the development of phylogenetic theory. Hennig (1966) corrected certain problems associated with the Biological Species Concept (see Meier and Willmann papers, this volume) but, regrettably, other equally severe problems remained. With the general adoption of phylogenetic theory in taxonomy, it became obvious that a concept of species was necessary that was consistent with phylogenetic thinking yet independent of cladistic analysis so that the elements for such studies might be assessed beforehand. At the same time, the emerging biodiversity crisis has made it imperative that we find better scientific ways to measure biological diversity (e.g., Nixon and Wheeler 1992b). One obvious need is for a unit species concept, so that the relative diversity of taxa (clades), areas, and habitats might reasonably be considered. From our point of view, none of the competing concepts can fulfill this need as completely or as precisely as the Phylogenetic Species Concept (as used in our position and critique papers).

A dominant focus on mechanisms of evolution, reinforced and expanded by the rise first of modern genetics and then of molecular biology, characterized most of the decades following Darwin's *Origin* (1859). This is unfortunate from the perspective of biodiversity studies because the core biodiversity sciences, especially taxonomy, have been neglected and inadequate support has existed to provide phylogenetic hypotheses and predictive classifications for most groups of organisms. In fact, it can be argued that the "biosystematics" of a few decades ago did considerable damage to taxonomy and delayed a number of important advances in our knowledge (Wheeler 1995c). One casualty of the New Synthesis was, from our point of view, the suppression of a taxonomic view of species by the genetic process-based Biological Species Concept, at least in many areas of zoology. All of the competing concepts, in one way or another, make certain ontological assumptions about evolution that unnecessarily constrain our notions of species before cladistic studies are undertaken. Although there are numerous theoretical assumptions underlying our Phylogenetic Species Concept (e.g., assumptions about homologies, traits, characters, hierarchical structure in data, etc.: see Platnick 1979; Nixon and Wheeler 1992a), it avoids additional unneeded ones.

This debate will hopefully fuel the kind of dialogue and argumentation among biologists necessary to resolve the so-called species problem, perhaps in time to provide a conceptual framework for responding to the biodiversity crisis. The use of the phrase *phylogenetic species concept* by us and by Mishler and Theriot presents a level of confusion both to the book and to the general discussion of species (Mayr, for example, either fails to bother or does not yet understand the distinctions between the two). We hope that this debate will help establish two facts in this regard: first, that our version of the Phylogenetic Species Concept is more useful in both a taxonomic and general biological sense and should be adopted over the other phylogenetic concept as well as the others in this book; and second, that it is both logical and descriptive to apply the name *phylogenetic* to our concept because of its consistency with phylogenetic theory and the recognition of elements of cladistic analysis. We would suggest that the Mishler and Theriot version be called instead the Monophyletic Species Concept or the Autapomorphic Species Concept, which, metaspecies notwithstanding, better describes their conceptual position.

Reply to Mayr

Most of Mayr's statements are platitudes that he has repeated on numerous occasions and that require even less response now than previously. As systematists who have published books and papers on far more species of organisms than has Mayr, we simply reject being labeled as "armchair taxonomists."

To our knowledge, no contributors to this volume have ever confused species as taxa with the species as a category. More importantly, Mayr's complaint that "several of the so-called species definitions did not define the category species at all" is silly, for as a category, there is no controversy whatever about the definition of *species*—indeed, it is the same under all species concepts (i.e., a subunit of a genus).

Mayr continues his complaint, however, by noting that "several of the so-called species definitions . . . were simply operational prescriptions as to how to delimit species taxa." Although it is true that our Phylogenetic Species Concept, by virtue of its reliance on diagnosability of samples, can be applied to the real world, it is not "simply" operational, and was not designed or chosen to meet the standards of operationalism. It was designed and chosen because it provides a basis for investigating phylogenetic relationships among the units it discriminates, without at the same time imposing arbitrary limits on the evolutionary processes that could conceivably have given rise to those units. In that respect, our concept actually seems intermediate in "operationality" between, on the one hand, the reductionistic approaches of Nelson (1989a) and Vrana and Wheeler (1992), which could apply as easily to individual organisms as to diagnosable samples, and, on the other hand, those concepts (such as that of Wiley and Mayden) that lack any obvious means of application to the real world.

Mayr also trots out his standard insult: "The proponents of the so-called Phylogenetic Species Concept quite openly return to the old typological species concept." Unless Mayr can provide an example of an "old typologist" who had a cladistic concept of characters rather than a phenetic one, this brush can carry no tar. "Nothing whatsoever is said about the biological meaning of the species concept." The biological meaning of phylogenetic species is precisely the same as that of Mayr's biological species, except in those cases where Mayr's concern for presumed reproductive compatibility has been allowed to override empirical evidence that "subspecies" or other samples improperly merged under one binomen are actually diagnosable. In those cases, of course, the biological meaning of phylogenetic species is far more obvious, and precise, than is that of the so-called biological species.

According to Mayr, a "further weakness of this concept is that it does not properly define what 'a unique combination of character states' is. For instance, owing to the advances in molecular biology, we can now determine unique character combinations for certain molecules characterizing human races. Would the Phylogenetic Species Concept necessitate making these races true species? Of course not. Then what is a unique combination of character states?"

This comment requires some unpacking. First, we are talking about samples, not individual organisms. Given the vast extent of interbreeding among current human populations, it is unlikely (although not impossible) that any geographically restricted sample of humans is diagnosable today, by sequence data or any other means. However, if we ask what the situation may have been prior to the advent of modern humans and their vastly increased mobility, it is certainly entirely possible that such diagnosable (and geographically discrete) populations did exist and could reasonably have been regarded, at that time, as separate species. That subsequent changes in human mobility have probably erased all remnants of this evidence of incipient speciation is just a historical contingency, not a theoretical problem.

Mayr's major complaint is thus that considering "every distinguishable geographic isolate as a separate species . . . would lead to a massive increase in the number of recognized species in all groups with geographic variation and isolation." In actuality, it appears that such a massive increase in the number of recognized species can be expected only in a relatively small number of groups, such as birds, in which systematists who have been misled by the Biological Species Concept have mistakenly reduced the number of recognized species over the past half century. In most highly speciose groups, such as insects and arachnids, no such massive increases are likely to occur, for most workers, from Linnaeus to date, have always used a species concept founded on diagnosability.

Mayr claims that "One of the major weaknesses is that it is left to the arbitrary judgment of the taxonomist what he or she considers to be such a smallest aggregation or what is diagnosably distinct." If by that criticism he means that what is today regarded by us as a single species might tomorrow be shown by someone else to be a complex of separate species that are each diagnosable, we stand guilty as charged and would have it no other way.

But it seems that what Mayr really fears here is something else: "When every diagnosable population is called a different species, the species of a genus will become exceedingly heterogeneous. Some of them are closely related (allopatric), and others are sympatric and only distantly related." In other words, a monophyletic group might actually contain some species that are widespread, occupying an entire continent, and others that are endemic to tiny areas. How very inconvenient of evolution to have produced such disparity!

Mayr imagines that our species concept is somehow connected with a search for stem species, but that criticism is misplaced and should be directed to Meier and Willmann instead.

Finally, Mayr reiterates his claim that the most profound criticism of our Phylogenetic Species Concept is that it has no biological meaning whatsoever. It is an arbitrary construct of the human mind. As a claim, these statements are fully as empty here as they were earlier in his critique. Both phylogenetic and biological species are constructs of the human mind in that they are hypotheses. The taxa placed as phylogenetic species have, in some cases, exactly the same biological meaning as do Mayr's biological species. In other cases, however, they have a far greater and far more precise biological meaning, for (unlike Mayr's) they do not purposefully (if foolishly) hide biodiversity from "the person who studies species in the field."

Reply to Mishler and Theriot

We respond first to Mishler and Theriot's "major basic issues" and then to their specific criticisms.

ONTOLOGY

Mishler and Theriot are entirely correct in separating us from the other authors in this regard. We do not believe that species taxa should in any way be based on "a priori theories about the speciation process" (sic; there are presumably many different speciation processes). Their ontology, manifest in part in the notion that species are individuals, has simply had no demonstrable impact on the discrimination or analysis of species. Provided that the distribution of attributes among samples (i.e., populations or sets of asexual organisms) is objectively assessed, belief in individualism of species has no more bearing on this science than does belief in UFOs.

SPECIAL REALITY

Mishler and Theriot are entirely incorrect in supposing that we differ from them in regarding "the reality of species . . . [as] somehow different" from that of higher taxa.

Monophyletic groups are just as real as species; indeed, we often have much better evidence for their existence, because not all species have detected autapomorphies. Certainly species occupy a unique position in cladistics, as the basic units of cladistic analysis and the hypothesized upper limit for tokogenetic relationships, but they remain, like higher taxa, just testable hypotheses based on characters.

UNIVERSALITY

Mishler and Theriot are correct in that we believe our species concept "can fit all groups of organisms." This is due, of course, to the fact that it is based on the empirical distribution of hypothesized characters (rather than upon assumptions about speciation processes) and the patterns shown by the diagnosable end-products of any and all evolutionary processes. For this reason, no modification of our Phylogenetic Species Concept is needed to deal with asexual species.

GROUPING VERSUS RANKING

We have no objection to Mishler and Theriot's characterization (table 9.1, rows 5 and 6) but do not concur that "all concepts avoid naming formal species taxa where they might be ephemeral or temporary (e.g., small, geographically isolated populations), even when they otherwise fit the criteria of the concept." That geographically isolated populations which are diagnosable today could conceivably lose their diagnosability in the future is no reason to ignore their present-day differences. Such an argument, if followed to its logical conclusion, would preclude recognition of any allopatric species.

DEFINITION OF MONOPHYLY

Mishler and Theriot are correct in that we use *monophyletic* and its variants as adjectives referring only to groups of species. Our usage is consistent with the formal original definition of the term and its intended use to describe phylogenetic and not tokogenetic or other kinds of relationships (Hennig 1966). The alternatives are unappealing, to say the least (or have I, as a monophyletic organism, just become paraphyletic because I sneezed?).

SEXUAL PROCESSES

We're for 'em. But Mishler and Theriot are correct that we give primacy to empirical evidence, such as diagnosability or the lack thereof, rather than to the supposed cohesion produced by sexual reproduction.

CHRONOSPECIATION

We regard this as a lame Felsensteinian exercise in model building. Because we live at one geological time plane, we study specimens at one time plane only, not as lineages whose behaviors are known. The ability of systems to somehow mirror the supposed behaviors of imaginary (i.e., modeled) lineages tells us little, if anything, about the real world and much about the limits of such models.

Most of Mishler and Theriot's specific comments concern asexual species; they see sexually reproducing organisms as having populations as evolutionary units, and asexually reproducing organisms as having individual organisms as evolutionary units, which may be reasonable enough, though insufficient. But there is nothing that dictates that the results of these two processes must produce significantly differing taxonomic patterns. In either case, there emerge diagnosable samples not (at least with current knowledge) further divisible into smaller such samples. There may be many such units in some groups of asexually reproducing organisms. But then again (at least in many of those groups), it may be important to recognize every possible such differentiate (consider, for example, the human immunodeficiency virus). Certainly, the claim that "asexuals . . . by definition have fixed characters" is preposterous, unless they mean merely "some fixed characters" (even asexually reproducing organisms are not always exact clones; if they were, no problems at the species level would ever arise because no confounding differences would ever be found).

Furthermore, by focusing on samples rather than individuals, our Phylogenetic Species Concept seeks to identify diagnosable sets of individuals (i.e., those that share a unique combination of attributes). This, of course, eliminates the necessity of recognizing as taxa individual organisms and their unique mutations. Mishler and Theriot confuse our interest in character distribution patterns that are interpreted as the results of various evolutionary processes with (their own) assumptions about those processes (i.e., that populations are units of such processes in sexual forms and individuals are the comparable units in asexual forms). The advantage of having the elements in a cladistic analysis be independent of such assumptions should be obvious.

Of interest, in this regard, is Mishler and Theriot's discussion of parthenogenetic species. Their invocation of metaphyly is required just by the fact that such species do not fit their view of hierarchy (i.e., mothers do not act like responsible, Hennigian stem species and die just because their parthenogenetic babies are born). But parthenogenetic species do sometimes occur within groups that are otherwise sexually reproducing, and they seem not to have provided any particular problem for our Phylogenetic Species Concept. For example, one species of an otherwise ordinary genus of ochyroceratid spiders is apparently parthenogenetic; it has attained a pantropical range (unlike its sexual congeners, which tend to be very narrowly distributed). Recognizing members of the parthenogenetic species has not proved to be a more difficult task than identifying other species in the genus. It remains possible that some of the various populations of this widespread species may prove to be diagnosable. The same proviso is true of the sexually reproducing members of the genus, of course. And from the biogeographic point of view, it would be enormously interesting if in fact some populations of the parthenogenetic species could be shown to be diagnosable, for those results would impinge significantly on our views of the history of the entire group, both in time and space.

But perhaps it is really just the stem species bugaboo that produces Mishler and

Theriot's waffling. How else is it possible that "cladistic analysis is absolutely necessary for determination of natural lineages of any level in asexual species"? The parthenogenetic spider species referred to above shows a unique combination of characters and was therefore successfully described long ago, without any efforts whatsoever at using cladistic analysis to determine natural lineages within its members. Indeed, the systematist who initially described the species was quite as unaware that it is parthenogenetic as he was of cladistic methods

Finally, we have little use for Mishler and Theriot's reliance on practical decisions regarding the "perceived importance" of groups. In actuality, such concerns may influence how much time, energy, and resources get devoted to studying a particular group, but promoting those concerns to theoretical significance in determining what taxa should be named seems entirely unjustified and inappropriate. Similarly, species are neither paraphyletic nor apomorphic. Definitionally (Hennig 1966) and conceptually, fluctuating frequencies of alternative genes make the notion of relative plesiomorphy and apomorphy of traits cladistically meaningless (Nixon and Wheeler 1992a).

Reply to Willmann and Meier

Willmann and Meier raise several criticisms that address both the philosophical underpinnings of our Phylogenetic Species Concept and implications for its application in practice. We respond to these concerns in the order in which they are raised.

PUTATIVE ANTIEVOLUTIONISM

There is no question that Hennig (1966) incorporated more evolutionary process assumptions into his positions than did his "transformed" followers (Platnick 1979). Yet the so-called transformed cladistics (also known as New York cladistics, pattern cladistics) is nothing more than the logical refinement of Hennig's arguments made more rigorous by the exclusion of unnecessary assumptions. As such, transformed cladistics is cladistics or Hennigian phylogenetic systematics much as evolutionary cladistics is a modified version of the evolutionary taxonomy that preceded it and the phylogenetic theory revolution.

Although cladists have argued against references to unnecessary evolutionary assumptions, few if any have advocated the abandonment of evolutionary thought entirely. Without the fundamental assumption that evolution has taken place and that there is a history of descent with modification, what justification would exist for ascribing significance to the hierarchical patterns reflected in cladograms? The ontological presumption that there is such a history and that hierarchic patterns are to be expected, however, falls far short of accepting a priori any particular evolutionary process (e.g., a preferred mode of speciation).

There are many additional theoretical assumptions embodied in our Phylogenetic Species Concept related to concepts of characters and traits, populations, and ancestral polymorphism (Nixon and Wheeler 1992a). Such assumptions are implicit in all scientific hypotheses and are not a particular issue unless they constrain hypothesis formation or testing in ways that are unnecessarily restrictive or theoretically unjustified.

Mayr suggests that our Phylogenetic Species Concept is far too operational, at the expense, in his view, of any biological content. Willmann and Meier, on the other hand, suggest that the fact that any evolutionary assumptions are made by our Phylogenetic Species Concept renders it less operational than claimed. We certainly hope that our Phylogenetic Species Concept is operational to the extent that it can be applied and critically tested in real-world circumstances; whether it is as operational as Willmann and Meier think we claimed is irrelevant, and we had no particular level of operationalism in mind. Any concept, however, that is divorced from the observable evidence necessary to critically evaluate its application should be rejected in favor of those that are open to potential falsification based on empirical observation.

By minimizing unnecessary assumptions, we do not become antievolutionary. To the contrary, we permit the recognition of the elements of cladistic analysis (i.e., species) so that a pattern may be established that is in need of evolutionary explanation. This and related transformed cladistic analytical methods are not only consistent with evolutionary theory, they are necessary if any particular evolutionary process is to be studied in a testable framework.

CHARACTERS AS HYPOTHESES

Willmann and Meier are correct that we regard characters to be hypotheses in the sense described by Nelson and Platnick (1981:301). Thus, when we say that characters are "observable," we mean that their varied original and subsequently modified states are observable and that their predicted combined distribution may be tested. Perhaps the point is made best by example.

Consider the presence of four ambulatory limbs in quadrupedal vertebrates. Among subsequent modifications of the forelimb are wings of birds and bats, flippers of penguins, and "leglessness" of snakes and certain lizards. Are the limbs of mice, penguins, falcons, bats, snakes, and limbless lizards immediately recognizable as homologues based on casual observation? Not necessarily so. Yet cladists make the prediction that all quadrupeds share the presence of four limbs, an assertion that may be tested by observing all species claimed to belong to the clade and an assertion that has been corroborated to date. This character does not fit the simple description of the pheneticist or the atheoretical construct that Willmann and Meier falsely ascribe to our Phylogenetic Species Concept. Cladistics, character analysis, and application of our Phylogenetic Species Concept are all rich in theory yet open to critical testing.

ARE SPECIES FOREVER?

Willmann and Meier appear to be bothered by the fact that a species may be claimed to exist at one time, then retracted and rejected at a future time. For us, species are simply hypotheses so that their acceptance or rejection relates to available evidence and the patterns revealed by its study. In such a context, it is neither surprising nor disturbing that a perfectly reasonable hypothesis at one time might be rejected given data at another time.

Reasons for such hypothesis revision are varied. Inadequate sampling of one or another population may have given a false impression of character transformation where undetected, perhaps low levels, of polymorphism actually exist. Temporarily allopatric populations may indeed have progressed to character transformation and speciation, only to introgress when a barrier breaks down at a future date and once again become polymorphic. At the earlier allopatric time, our Phylogenetic Species Concept would predict that they were species and this would be fully theoretically justified. At the later sympatric and polymorphic stages in their history, our Phylogenetic Species Concept would hypothesize that they were two populations of one species. Each would represent the best possible hypothesis at that point in time. We do not regard species to be immutable, but merely the most reasonable hypothesis given existing data.

IS GENE FLOW PERMISSIBLE BETWEEN SPECIES UNDER OUR PHYLOGENETIC SPECIES CONCEPT?

Yes. The existence of hybrids is not unexpected based on what population geneticists have learned and does not negate the existence of either parent species. Willmann and Meier imply that the existence of hybrids falsifies the claim of character constancy in the parent species and therefore undermines our Phylogenetic Species Concept.

Under our Phylogenetic Species Concept, we would hypothesize that character distributions are constant within each of the parent species. If a significant introgressive genetic situation exists, it seems unlikely that such intrapopulational constancy would long be maintained. Hybrids are just that—hybrids—and not self-perpetuating kinds in the same sense (including character distribution sense) of either parent. Their existence does not negate the existence of their parents any more than a solar eclipse negates the notion of day and night.

HOMINIDS, PHYLOGENETIC SPECIES, AND EVOLUTIONARY END-PRODUCTS

Willmann and Meier clearly misunderstand the phrase *evolutionary end products.* Descent with modification is an ongoing and unending quality of life on earth, and in that literal sense there can never truly be end products of processes that do not cease to exist. A reasonable interpretation of *end product* for the phylogeneticist is thus to analyze the distribution of characters among organisms at a particular point in time, often the present. This is why, in part, it is desirable to have a species concept that is character based. It provides a framework for understanding and predicting character distributions among populations regardless of whether those distributions were the same or different in the past.

"According to the Phylogenetic Species Concept of Wheeler and Platnick, the Australian aborigines were definitely a separate species before the invasion of Caucasians" (Willmann and Meier, critique paper). As we stated above, prior to the advent of intercontinental travel in the past few hundred years, it does appear probable that character distributions would have suggested more than one species of human on the planet. Unlike Meier and Willmann's concept, however, our Phylogenetic Species Concept does not claim immutability, and the obvious contemporary pattern of increasing introgression among previously allopatric human populations suggests that we are or soon will become one global polymorphic species. One purpose of a species concept is to assess precisely how many populations have completed character transformation and attained a distinctness sufficient to support the prediction of character distributions among populations. Such units are what we call end-products. In the case of humans, what would have been properly hypothesized as a number of end-products a thousand years ago is today most reasonably hypothesized as a single one. This only serves to elevate the relative contribution of arthropods to our planet's biodiversity and emphasize the plasticity of evolution.

QUANTITATIVE CHARACTERS AND OUR PHYLOGENETIC SPECIES CONCEPT

Willmann and Meier suggest that quantitative characters are problematic in a theoretical context where constancy of character distributions is a requirement. We respond with two simple observations. First, characters are hypotheses and not simple phenetic observations, as discussed above. The way in which a character is conceived and described determines whether it is useful or not in a species delimitation or cladistic analysis connection. And second, quantitative characters are a problem of character analysis shared by all systematists regardless of their preferred species concept. Such evidence is problematic and its analysis is frequently challenging, but this is not a challenge unique to the Phylogenetic Species Concept.

ANAGENETIC SPECIES AND OUR PHYLOGENETIC SPECIES CONCEPT

Willmann and Meier would have us recognize cladogenesis without character transformation on the one hand and reject populations demonstrating complete character transformation when no cladogenesis is thought to exist, on the other. Because our Phylogenetic Species Concept rests upon hypothesized characters, we are led to recognize species regardless of how they may or may not have formed. Before character transformation, there exists no unambiguous way to know that two isolated populations are destined to become distinct species at some time in the future or which might simply reunite. Thus, recognizing cladogenesis without character transformation is unfounded. Because we know of no way to definitively say whether a species arose by allopatry or sympatry or by this or that mechanism of speciation, we seek a concept of species that applies to all conceivable combinations of evolutionary modes and factors. Anagenesis looks the same in practice as cladogenesis with extinction of the ancestral form, and each character in a chain of anagenesis might well correspond to an undetected cladogenetic event. For our Phylogenetic Species Concept, such characters do mark significant evolutionary events and point to new end-products distinct from their ancestral populations. By permitting the observed distribution of hypothesized characters, we can discover evolutionary patterns without arbitrarily shaping them based on our presumptions about possible causal factors.

Reply To Wiley and Mayden

Ancestors

We agree with Wiley and Mayden that ancestral populations may be present in a cladistic analysis and that there is no theoretical reason to expect such ancestors to have even a single apomorphy not shared by all the descendant species. However, we fear that another aspect of our Phylogenetic Species Concept was not presented with sufficient clarity. When gene frequencies shift in a polymorphic ancestral species, none of the alternative allelic combinations is meaningfully described as apomorphic or plesiomorphic relative to other infraspecific traits. Stated another way, the beginning condition is polymorphism, from which any of the alternative traits could potentially become fixed.

Cladogenesis versus Character Evolution

Like Meier and Willmann, Wiley and Mayden ascribe significance to cladogenesis prior to demonstrated character transformation. No doubt populations of widespread species can become isolated for brief periods of time, only to subsequently reunite in secondary panmixis. There is no way to know, when allopatry occurs, whether it will ultimately result in character transformation or speciation or whether it will return to panmixis in any particular case. Although it is hypothetically possible to draw such conclusions retroactively, rarely do data support such conclusions, and never unequivocally, and there is never justification for predicting what currently isolated populations will become in the future except where character transformation has already taken place. In such cases, our Phylogenetic Species Concept can predict that a population is a separate species and that it will retain character distinctness; this prediction is potentially falsified by additional observations of character distributions.

Rates of Evolution

Sister species may be treated for all intents and purposes as having the same age. That age is set as the point in time at which character transformation was complete and speciation established, not the earlier time at which initial cladogenesis (generally allopatric isolation) was set into motion. Only by so doing can we stick to the facts at hand.

Wiley and Mayden are correct that different transformation series may progress at different rates, hence the traditional descriptions of relative tachytelic and bradytelic evolution. The issue of speciation without cladogenesis was already addressed above.

ARE CHARACTERS HIERARCHICAL?

Wiley and Mayden claim that characters cannot and do not form true hierarchies. As attributes of organisms or, more precisely, of semaphoronts (Hennig 1966), characters do not exist disembodied from living individuals. The cladistic hypothetical characters, however, do fulfill the logical requirements of a strict hierarchy (Woodger 1937). They have a unique beginner, the original ancestral attribute. Through subsequent modifications of that original character they progress unidirectionally from one to many. And such characters form nested sets of increasingly inclusive composition, just as sets of species may be grouped into nested clades (Platnick 1979). The fact that they meet these requirements means that we can analyze their relationships to one another and reasonably draw historically significant conclusions from their patterns of relationship. Were this not so, then the entire enterprise of phylogeny could not exist. Protestations of ontologists like De Queiroz (1988) notwithstanding, we have no way of discovering evolutionarily historical patterns without reference to characters. Ontology ungoverned by a logical epistemological basis for the study of species and clades leads to a belief system that does not differ significantly from creationism.

INDIVIDUALS VERSUS CLASSES AND CHARACTERS VERSUS TRAITS

The philosophical distinction between individuals and classes is not so stark as Wiley and Mayden would have us believe, and we tire of the incessant reference to individuals. Species, for us, are the smallest elements of cladistic analysis that cannot be divided into smaller units for which unambiguous information about phylogeny exists. They are indeed grouped together into classes (i.e., clades) based on synapomorphy.

Wiley and Mayden confuse character and trait in their strict sense (Nixon and Wheeler 1992a). A trait is an attribute that varies among individuals or populations within a single species and is the thing studied by population biologists. A character is an attribute that varies between but not within species and as such provides potential evidence of shared ancestry among species. Unless a clear distinction is made between traits and characters, all kinds of erroneous conclusions can be drawn, not the least of which are the importation of probabilistic concepts from population genetics to phylogenetics and the supposed cladistic analysis of populations within a single species.

Wiley and Mayden argue that characters do not give rise to other characters. We are as aware of the biological role of egg bearer and sperm donor as Wiley and Mayden, but regard their rejection of characters giving rise to characters as little more than a semantic sleight of hand. As explained by Platnick (1979), all descendants of a common ancestor share its heritable characters either in its original or in some subsequently modified form. This conceptual linkage of characters through transformation is fundamental to cladistic theory. Characters are not spontaneously generated and are always modifications of preexisting attributes of an organism. It is far more scientifically profitable to view characters as giving rise to transformed characters than to get hung up on mechanisms of evolution. The latter are too numerous and act in combinations too unpredictable to offer any meaningful conceptual basis for understanding historical patterns.

Characters, narrowly defined based on their infraspecific constancy, are an emergent property of species and of clades. The fact that character bearers are individual organisms is a biological fact but a reductionist thought irrelevant to the discovery, study, or critical testing of characters or their distribution.

16

A Defense of the Evolutionary Species Concept

E. O. Wiley and Richard L. Mayden

We have organized our responses first on an author-by-author basis. In some cases, a single topic is discussed in more than one section so that we can respond directly to particular criticisms. We present a summary after dealing with each of the critiques of the Evolutionary Species Concept.

Reply to Mayr

TAXA VERSUS CATEGORIES

We are not sure if Mayr is criticizing us for not making the distinction between taxa and categories. *Fundulus lineolatus* is a taxon. Because it is hypothesized to be a lineage and not a monophyletic group of lineages, it is given a binominal according to the rules of naming taxa of the species category. If *F. lineolatus* is later found to comprise two lineages, then we would apply two binominals, each formed according to the rules of naming taxa of the species category. If *F. lineolatus* is found to be part of a larger lineage, then we would combine names according to the rules of priority and adopt a single binominal.

BASIC OBJECTIVES OF BIODIVERSITY

We see the basic objectives in biodiversity research today as threefold. First, inventory (e.g., discover and describe) biological diversity. Second, conserve and manage biodiversity resources. Third, educate the public as to the significance of biodiversity, and hence, the significance of systematics and taxonomy. Appreciation can only come through education.

In his book on biodiversity, Wilson (1992:37–38) argued that the eventual success of the field of biodiversity is inextricably linked to the discovery of a basic "atomic unit" with which biodiversity "can be broken apart, then described, measured, and reassembled. So, the species concept is crucial to the study of biodiversity. It is the grail of systematic biology. Not to have a natural unit such as the species would be to abandon a large part of biology into free fall, all the way from the ecosystem down to the organism." A similar view is shared by Paterson (1993:137): "When we trade in ideas in population biology, species constitute the currency." The bottom line is this: whether they know it or not, myriad research programs, institutions, agencies, educators, and people in general are heavily dependent on the topic of species and an enormous number of related topics and fields because the species is the atomic unit in biodiversity.

We submit that the Evolutionary Species Concept is superior to the Biological Species Concept as an atomic unit for biodiversity. The Biological Species Concept advocates a chauvinistic perception of diversity, one obviously in discord with known, natural, biological systems. Given that all asexual species are disregarded and that most allopatric lineages, regardless of their sexual tendencies, are only considered subspecies, biodiversity recognized under this concept is severely abridged. Adding to this human-induced imposition on nature, any taxon breeding outside its species boundary, regardless of the reason or its genealogical relationship to a potential mate or the ultimate impact of this transgression, will be subsumed into a black hole of terminal taxonomic neglect and ambiguity, better known as the subspecies category. This is a serious problem for everyone because the vast majority of cladogenesis occurs in allopatry and a significant segment of the diversity spectrum reproduces without sex. The Evolutionary Species Concept has no such limitation. Debate as to what constitutes a particular species can still occur under the Evolutionary Species Concept, but that debate will be focused on real issues at the population level (relative isolation, critical habitat demands) and not at the superficial level of opinions regarding propensity to interbreed if sympatric. Recognized units of biodiversity under the Evolutionary Species Concept will not be recognized and conserved just by their predilection for having sex. Diversity will not be lost just because a geographic population is considered a mere subspecies whose numbers can simply be replenished from some other area of the species range. In short, we submit that the Biological Species Concept underestimates true biodiversity at the species level and that the Evolutionary Species Concept provides a more useful set of criteria to estimate this diversity.

SIMPSON'S WEAKNESS

Whether the Evolutionary Species Concept of Wiley (1978) suffers from the same weaknesses as Simpson's (1961) or shares the same strengths as Simpson's (1961) is a matter of perspective. Virtue to the grasshopper may be vice to the ant. We, in fact, have pointed out that Simpson (1961) advocated one concept (species as lineages) but practiced another (recognition of phyletic species for taxonomic convenience [Wiley 1978; see also our position paper]). Every isolated population is expected to become, or has the potential to become, a new species. Of course, what constitutes the quality of *isolated* is not straightforward. As Lewontin (1974:213) demonstrated, gene flow as infrequent as one migrant per 1,000 generations may prevent effective isolation. What may appear to be an isolated population during our life span may, in fact, not be isolated at all. We assume that character differences constitute empirical evidence that can be used to recognize isolated populations. Certainly, if there are no character differences, then we would not even suspect that a population was isolated. And certainly, the quality of the differences might suggest whether isolation was "practically complete" (more than one species suspected) or only suggestive of geographic or ecophenotypic variation. In this respect we feel that we are conventional systematic ichthyologists, applying the same empirical evidence as many of our ichthyological colleagues who do not share our passion for theory. We have never claimed to have special or mystical knowledge that is unavailable to others. The proposition is straightforward. Gene flow between populations tends to erase differences between those populations (standard population genetic theory). A reasonable person can conclude that a lack of gene flow leads to differentiation and that a complete lack of gene flow leads to diagnostic differences. Lack of gene flow (directly or indirectly measured) appears to us to be reasonable evidence of lineage independence. If so, this constitutes empirical evidence that we are dealing with evolutionary species and not populations. The quality of lineage independence, then, is not mysterious. It is derived from standard population genetic considerations coupled with macroscopic considerations of the effects of vicariance and other higher-level processes on gene flow.

Under the Evolutionary Species Concept, when Professor Mayr names a subspecies of bird he is making a specific evolutionary statement, to wit: this is a partly differentiated lineage of birds that is still in genetic contact with other populations of the species. When we name a species of fish under the Evolutionary Species Concept, we are also making a specific evolutionary statement, to wit: these fish comprise a separate evolutionary lineage within which character change is decoupled from its closest relatives because gene flow has ceased to be the deciding factor in the evolutionary history of the species pair. We see no operational problem with our definition. It leads to specific testable consequences.

Reply to Mishler and Theriot

TOKOGENY

It seems that Mishler and Theriot are the only ones who got Hennig right about tokogeny. All the rest of us got it wrong. Actually, Mishler and Theriot seem to have misread our paper. We state that "parts of species (individual organisms) participate in tokogeny." If asexual species are evolutionary species, a position that we advocate, then how can we be criticized for equating tokogeny only with sexual reproduction?

EPISTEMOLOGY

Yes, we do claim that there is no simple epistemological approach to discovering an evolutionary species. We believe, though, that there are several direct approaches: study biogeography, study morphology, observe breeding behavior, study gene flow or lack thereof. What is the problem?

ANCESTRAL SPECIES

The question is put: "[I]f tokogenetic continuity is maintained and lineage splitting does not demarcate the end of a lineage, then what does?" Extinction marks the end of many lineages. And, we suspect that in the vast majority of cases what comes out of a speciation event is two or more new species and that ancestral species do, in fact, go extinct as tokogenetic entities. We have never said otherwise. This is a prediction derived from some fairly simple population genetics coupled with the empirical observation that vicariant speciation (model I of Wiley 1981a) seems to be a common or even predominant mode of speciation (Lynch 1989). The analogy with other individuals is simple: an amoeba that splits gives rise to two new daughters. But for speciation via hybridization (or reticulation, if you will) and esoteric bouts of peripheral isolation, perhaps another analogy can be drawn, the analogy of the father and mother who do not have to change their names when their daughter is born or the analogy of the hydra that buds a daughter. Because a central concept of the Evolutionary Species Concept is that species are individuals, we simply would like to allow species to minimally have the capacities and characteristics of other sorts of individuals. When some individuals reproduce, they become extinct. When other individuals reproduce they do not become extinct until some time after the act of reproduction. Maybe ancestral species must become extinct. If so, it is not the fault of the concept of species as lineages but the fault of Wiley (1978, 1981a) and Wiley and Mayden (see our position paper) for their cognitive deficiencies. We suspect that species do not have to become extinct but will be happy to consider why the origin of any particular agamospecies via a hybridization event should cause the lineages of each parental plant to be renamed, with the original name being applied to all specimens before the event and the new names being applied to all specimens after the event. If the ancestors of species arising via hybridization can survive the speciation event, then there can be no absolute logical reason to preclude ancestors from surviving speciation events. We present additional discussion of this subject in the Willmann and Meier reply.

ASEXUAL SPECIES

Yes, we do think that there are ontological differences between purely asexual species and purely sexual species. We also think that there are some ontological similarities, just as we see the ontological similarities between sexual species, asexual species, and higher taxa. The similarities of asexual and sexual species as tokogenetic entities make them candidates for being included within the same broad species concept, the Evolutionary Species Concept. This may be wrong, but we hardly think that it is incoherent. We are sorry Mishler and Theriot missed this point. We would have liked to read their reaction. As to our inability to "defend asexual species," we simply do not know what Mishler and Theriot mean, so we have no defense.

Reply to Wheeler and Platnick

COMPATIBILITY WITH PROCESS THEORIES

We should certainly hope that the Evolutionary Species Concept is compatible with any and all process theories that we find meritorious. It would be a huge mistake to advocate a concept that flies in the face of well-established process theories. We find it curious that Wheeler and Platnick would criticize us on this point but claim that this very same quality is a virtue of the Phylogenetic Species Concept. We suggest that Wheeler and Platnick cannot have it both ways. But, not to worry, the Evolutionary Species Concept is incompatible with some supposed evolutionary patterns and with claims about the nature of, or importance of, certain processes. Here are four major areas of incompatibility.

1. The Evolutionary Species Concept is incompatible with the process of phyletic speciation. It prohibits both the process and the pattern as basic violations of the quality of individuality embodied in the concept of species as lineages. In saying this, we do not imply that the Evolutionary Species Concept is incompatible with anagenesis. It is not. Character-based concepts, such as the Phylogenetic Species Concept *sensu* Wheeler and Platnick, do not prohibit phyletic speciation and thus should be rejected. We suggest that character-based concepts are not compatible with process and pattern because the application of character-based concepts results in patterns that are not produced in nature. In short, the Phylogenetic Species Concept applied to an evolving unitary lineage, with its subsequent multitude of branches, can overestimate the number of actual cladogenetic events that has occurred. There are at least two options for recourse. First, one could stack the names up on a single branch, similar to patterns produced in stratophenetics. Second, one could claim that the resulting branching diagram is irrelevant to real processes such as cladogenesis. But if this is the case, we fail to see the value of the exercise and how this concept meets the needs of phylogenetic systematics.

2. The Evolutionary Species Concept casts doubt both on the centrality of the acquisition of reproductive isolation and on the centrality of adaptation to the various processes of speciation. If speciation is a primarily allopatric phenomenon, then acquiring some form of reproductive isolation cannot be central to speciation because closest relatives are not sympatric. We do not, however, doubt that the acquisition of reproductive isolation has important macroscopic consequences. Effective reproduc-

tive isolation seems necessary if sympatry occurs between sister species subsequent to their origin. But this is one step removed from the origin of most species whose patterns of origin are allopatric rather that para- or sympatric.

3. It is an empirical observation that rates of speciation exceed (internal parasites, Brooks 1985) or far exceed (insects, Ross 1972) rates of ecological change and adaptation. If detectable adaptive and/or ecological changes lag far behind the rate of speciation, then adaptation to changing environments cannot be the primary force driving the production of new species.

4. It is an empirical observation that rates of character change exceed rates of cladogenesis (see any phylogeny on the species level and check the tick marks on the tree). If so, then character change cannot be driving cladogenesis. If all nonreticulate speciation involves cladogenesis, then the primary mechanisms involved in character fixation are anagenetic rather than cladogenetic, even when anagenesis is time compressed due to small population size.

HENNIG AND MORPHOLOGY

Wheeler and Platnick imply that Hennig (1966) did not repudiate character-based definitions of species as typological. But he did do so, and claiming otherwise is sterile revisionism. After an extended discussion ranging over 10 pages, Hennig (1966:43) concluded as follows: "There is neither a particular degree nor a particular kind of correspondence that justified regarding two semaphoronts in the phylogenetic system as members of one and the same group category of lower rank. Neither degree nor kind of similarity relationships between two semaphoronts permits conclusions as to the genetic relationships between the semaphoronts." Hennig (1966:43–44) concluded that comparative holomorphology can be used to "yield probabilities" and stated, "Thus the rejection of a morphological definition of the species does not mean that in practice systematics has to do without morphological aids in determining the limits of species."

We see no difference in our position on this point and Hennig's position. But it is clear that Hennig did indeed reject character-based species concepts such as the Phylogenetic Species Concept *sensu* Wheeler and Platnick. Later, Hennig (1966:78–80) made it clear that purely morphological systems are systems that he considers typological. We can only agree.

SPECIES AND ONTOLOGY

Wiley and Mayden do indeed contend that monophyletic groups and species do not exist because of synapomorphies. This would be like saying that Ed Wiley exists because he has a bald head or green eyes. Some monophyletic groups are discovered because we discover synapomorphies, and some species are discovered because we discover autapomorphies or diagnostic combinations of characters. But Wiley and Mayden do not know enough about the evolutionary process or about evolutionary history to claim that character transformation and speciation are inseparable outcomes of one and the same process. Such a conclusion could be drawn only if we decided to impose evolutionary rules on organisms in such a manner as to decide the outcome of our investigations a priori. We take a more moderate position. We take the terms *anagenesis* and *cladogenesis* as terms for two general kinds of processes, and we do not demand that they operate in a lock-step manner. Thus, we allow that some anagenesis leading to character fixation is not coupled with cladogenesis, and we allow that some cladogenesis might not be conveniently accompanied by some form of character fixation through anagenesis. We feel that our position on both points has been empirically demonstrated (see discussion above). We expect that some monophyletic groups and some species might not be discoverable using the analytical tools we have at hand (character analysis, biogeographic analysis, etc.). This is a pity, but every empirical research program has its limits, even phylogenetics.

We agree that the study of character variation does not deny ontology. But our claim, first articulated by Wiley (1981b), that characters are neither necessary nor sufficient to define (diagnose) taxa is not an ontological claim. It is an empirical claim deduced from the observation that an individual belonging to a species or monophyletic group may die before it develops the diagnostic character(s) of its taxon or because it lacked the DNA code to produce the character in question. We recognize the ontological claims of Wheeler and Platnick for what they are: species as taxa are classes. We prefer an ontology, species as individuals, that permits species to evolve rather than an ontology that requires specimens to be examples of timeless and immutable sets of characters. Perhaps species as classes would be an acceptable ontology if characters formed hierarchies, but we have rejected this notion (see our position paper) and thus opt for the logically sound alternative ontology that recognizes the place of characters in the phylogenetic system as aids in elucidating the common ancestry relationships of taxa (Hennig 1966).

Reply to Willmann and Meier

AMBIGUITIES

Willmann and Meier criticize the Evolutionary Species Concept as being ambiguous because useful concepts must include a criterion for delimiting species in the time dimension. Apparently this stems from the fact that ancestral species might give rise to descendants without becoming extinct. Willmann and Meier take us to task for asserting that Hennig (1966) would allow ancestral species to survive speciation events. We make reference to Hennig's *lapsus* on this point in our critique paper but will summarize our argument here. Hennig (1966:61) discussed the case of the New Zealand gall wasp, *Stenodiplosis geniculati* "var. *dactylidis*." This variety is diagnosable relative to the original European form and invaded New Zealand after the original description of *S. geniculati*. Hennig stated that if we considered the New Zealand form as a species, then "obviously no systematist would be prepared, on the basis of such reflections, to give another name to the species that is still called *Stenodiplosis geniculati*" (1966:61). So long as ancestors are abstractions, then it is fine to apply theoretical ideas such as requiring that absolutely all ancestors become extinct at speciation events. But, in the actual act of applying binominals, Hennig (1966) did exactly what Willmann and Meier prohibit: he allowed an ancestor to survive a speciation event.

It would seem that Hennig-the-theoretician differed from Hennig-the-working-systematist. That is not unusual. However, there is a deeper question. Is there a logical reason to insist on ancestral extinctions at speciation events? Frost and Kluge (1994) claim that there is and that insisting on ancestral extinction solves certain paradoxes that might exist involving the transition from tokogenetic to phylogenetic relationships (in their words, transitions between scalar and specification hierarchies). There is no logical prohibition placed on an ancestor to become extinct under the Evolutionary Species Concept. But such a prohibition might emerge from higher-level considerations of the kind discussed by Frost and Kluge (1994). In particular, it is obvious to us that one major conceptual difference between our attitude and the discussion of Frost and Kluge (1994) is that we view species as replicators, while they claim that species do not have this quality. If Frost and Kluge (1994) are correct, then evolutionary species may have to go extinct at speciation events as a matter of logic. We await further developments from that front. We note, however, that whether or not ancestral species can coexist with one or more descendants was no reason for Frost and Kluge (1994) to reject the Evolutionary Species Concept. In fact, they adopt it. It would seem that the question of ancestral extinctions has little or nothing necessarily to do with the acceptance or rejection of the Evolutionary Species Concept.

GENETIC ISOLATION

Willmann and Meier criticize the Evolutionary Species Concept because it does not adopt an absolute genetic isolation criterion. We believe their position differs from ours because they believe that tokogenetic events are synonymous with tokogenetic systems. We disagree. We hold that two evolutionary species may have members that occasionally hybridize and that may exchange genes through backcrossing. Such hybridization events constitute tokogenetic events between two tokogenetic systems. We submit that such events do not necessarily destroy the two independent tokogenetic systems. If occasional hybridization meant that we were dealing with a single tokogenetic system (a single species), then how could we continue to identify members of each of the parent species and how could we continue to identify hybrids and backcrossed individuals? (This exact point was made to E.O.W. by Donn E. Rosen, to whom credit is due.) As ichthyologists, we are aware of gene introgression among distantly related (nonsister) species. For example, individuals of the sunfish species *Lepomis cyanellus* and *L. megalotis* commonly hybridize, especially in marginal and disturbed habitats. Backcrossing is common in these situations. But this does not mean to us that *L. cyanellus* and *L. megalotis* comprise a single tokogenetic system. The hybrids are, if you will, tokogenetic outliers. They tend to breed back to one of the other species during subsequent generations. Such phenomena are common among members of *Lepomis* and many other fish groups. If we equate tokogenetic outliers with the limits of tokogenetic systems, then the total number of freshwater fish species will be drastically reduced.

If we are interested in Hennig's opinion, it would seem that he did not require absolute reproductive isolation. Rather, Hennig (1966:68) required "practically complete" reproductive isolation. We promise to abide by the "practically complete" criterion for sympatric species, although we realize that such a criterion will always yield ambiguities. We submit that Willmann and Meier are wrong to claim that evolutionary species are smaller than the largest tokogenetic systems because we do not think that tokogenetic outliers define the limits of such systems.

AGAMOSPECIES

Willmann and Meier briefly criticize our inclusion of agamospecies within the Evolutionary Species Concept. This is certainly a contentious issue, even among those who like the concept (cf. Frost and Hillis 1990). We agree that hierarchical relationships can be translated into classifications unambiguously only by creating nested taxa (i.e., by discovering that hierarchical relationships exist). But so what? Certainly some groups of agamospecies will display little or no hierarchical structure but rather polytomies. Perhaps others will display hierarchical structure for any number of reasons. When hierarchical structure is empirically demonstrated, then nested classifications will result. When we fail to find hierarchical structure, then hierarchical structure should not be imposed and we can fall back on terms such as *sedis mutabilis* (Wiley 1979b) to show our ignorance. Apparently Willmann and Meier feel that since descendant species cannot be part of their own ancestor, this creates a particular problem for agamospecies that is not present in bisexual species. But consider this: the ancestral species of all vertebrates is phylogenetically equivalent to Vertebrata itself (Hennig 1966). It was not "part of" Vertebrata. It was Vertebrata, and it stands logically equivalent to Vertebrata relative to its descendants (Hennig 1966:71–72; Wiley 1979b, 1981a). In the tokogenetic/phylogenetic sense, the ancestral vertebrate is not much different than any other ancestral species, whether that species is an agamospecies or a bisexual species. We can classify agamospecies exactly to the extent that we can diagnose them and elucidate, or fail to elucidate, hierarchical structure within the group to which they belong. We can do exactly the same thing with ancestors, persistent ancestors, or even taxa of hybrid or symbiotic origin along with their ancestors (Wiley 1979c). There is no problem here.

WILLMANN AND MEIER'S QUESTIONS

Willmann and Meier conclude with a number of questions. We answer them below.

1. Yes, absolute reproductive isolation does exist between groups of biparental organisms. Sometimes these groups are sister species and sometimes they are not sister species.

2. We would call units separated by reproductive isolation *taxa*. In some cases these taxa might be ranked as species within a larger taxon, such as one ranked as a genus (a suitably monophyletic genus). In other cases we might say that in our experience, all of the species of a monophyletic genus, family, order, and so forth, are reproductively isolated from all the species of another such taxon.

3. If speciation is predominantly allopatric, then we do not think that the origin of a new reproductive gap is directly related to the origin of a new species. There is no "need" to develop reproductive isolation if your closest relatives are in the next valley and you never encounter them. We would expect that if two sister species are separated for a sufficiently long period of time, then the chances that they have diverged sufficiently to develop reproductive isolation increases with that time period. This is a basic entropy argument applied to changes in genetic information (Brooks and Wiley 1988) and fully compatible with standard population genetic theory. But we must remember that the reproductive isolation did not evolve in order for speciation to be "completed." Of course, we can also envision immediate and complete reproductive isolation (e.g., speciation via hybridization resulting in polyplody), and we can entertain the idea that sympatric speciation requires this phenomenon.

4. We can imagine that the origin of reproductive isolation between lineages contributes significantly to the fate of previously geographically isolated populations that have become sympatric. The "cleanest" scenario is sympatry accompanied by total reproductive isolation because F^1 hybrids would never occur or would be infertile. But being practically reproductively isolated (Hennig 1966) seems to do the trick in fishes, allowing for sympatry without complete (100%) reproductive closure. Obviously, there is a correlation between the ability to interbreed and remoteness of relationships (C. L. Hubbs 1955; C. Hubbs 1967; McAllister and Coad 1978). This implies that reproductive isolation is as much due to time of divergence from the common ancestor as to the "need" to develop reproductive isolation. Unfortunately, as Rosen (1978) pointed out, the ability to interbreed is a plesiomorphic character and thus does not provide a metric for determining common ancestry relationships (contra McAllister and Coad 1978). But a loose correlation does exist, and that correlation is not itself that well correlated with the history of cladogenesis (e.g., see Rosen 1978), although it is apparently better correlated with cladogenesis than with the simple phenetic metric outlined by McAllister and Coad (1978). Put more simply, the hierarchy of integrated tokogenetic entities (lineages) is not completely correlated with the relative frequency of complete reproductive closure. Thus, we do not find Willmann and Meier's arguments compelling.

We advocate the Evolutionary Species Concept because it provides a conceptual link between pattern and process. We agree with Frost and Kluge (1994) that this concept maximizes the explanatory hypotheses of the phylogenetic system by providing a conceptual framework for binominals such that these binominals are associated with the largest integrating lineage (Wiley 1978). We reject the idea put forth by Willmann and Meier that these largest integrating lineages are defined by occasional or even common hybridization for the simple reason that such reproductive activity does not result in integration. Because species are not characters, we acknowledge that we will not discover all of them, and we acknowledge that binominals stand as hypotheses always subject to further testing as our knowledge base increases. We believe that definitions of concepts that do not arbitrarily limit us to what we can discover about species (see Frost and Kluge 1994) are superior to those that impose such limitations.

REFERENCES

Agassiz, L. 1854. Notice of a collection of fishes from the southern bend of the Tennessee River in the State of Alabama. *American Journal of Science and Arts* 2:353–369.

American Museum of Natural History. 1999. *The Global Taxonomy Initiative: Using Systematic Inventories to Meet Country and Regional Needs.* New York: Center for Biodiversity and Conservation, American Museum of Natural History.

Arnold, M. L., C. M. Buckner, and J. J. Robinson. 1991. Pollen-mediated introgression and hybrid speciation in Louisiana irises. *Proceedings of the National Academy of Sciences, U.S.A.* 88:1398–1402.

Arnold, M. L., J. J. Robinson, C. M. Buckner, and B. D. Bennet. 1992. Pollen dispersal and interspecific gene flow in Louisiana irises. *Heredity* 68:399–404.

Ashlock, P. 1971. Monophyly and associated terms. *Systematic Zoology* 20:63–69.

Avise, J. C. and R. M. Ball. 1990. Principles of genealogical concordance in species concepts and biological taxonomy. In: D. Futuyma and J. Antonovics, eds. *Oxford Surveys in Evolutionary Biology,* vol. 7, pp. 45–67. Oxford: Oxford University Press.

Ax, P. 1987. *The Phylogenetic System. The Systematization of Organisms on the Basis of Their Phylogenesis.* New York: Wiley-Interscience.

Bailey, R. M. (chairman of the Committee on Names of Fishes). 1970. *A List of Common and Scientific Names of Fishes from the United States and Canada.* 3d ed. Special publication no. 6. Washington, DC: American Fisheries Society.

Barr, T. C. Jr. 1979. Revision of American *Trechus* (Coleoptera: Carabidae). *Brimleyana* 2:29–75.

Barrowclough, G. F. and N. R. Flesness. 1996. Species, subspecies, and races: the problem of units of management in conservation. In: D. G. Kleiman, M. E. Allen, K. V. Thompson, S. Lumpkin, and H. Harris, eds. *Wild Mammals in Captivity,* pp. 247–254. Chicago: University of Chicago Press.

Baum, D. 1992. Phylogenetic species concepts. *Trends in Ecology and Evolution* 7:1–2.

Bell, M. A. 1979. Persistence of ancestral-sister species. *Systematic Zoology* 28:85–88.

Bock, W. 1979. The synthetic explanation of macro-evolutionary change—a reductionistic approach. *Bulletin of the Carnegie Museum of Natural History* 13:20–69.

Bock, W. 1995. The species concept versus the species taxon: their roles in biodiversity analyses and conservation. In: R. Arai, M. Kato, and Y. Doi, eds. *Biodiversity and Evolution*, pp. 47–72. Tokyo: National Science Museum Foundation

Bonde, N. 1977. Cladistic classification as applied to vertebrates. In: M. Hecht, P. Goody, and B. Hecht, eds. *Major Patterns in Vertebrate Evolution*, pp. 741–804. London: Plenum.

Bonde, N. 1981. Problems of species concepts in palaeontology. In: J. Martinell, ed. *International Symposium on "Concept and Method in Paleontology,"* pp. 19–34. Barcelona: Universitat de Barcelona.

Brandon, R. N. 1990. *Adaptation and Environment*. Princeton, NJ: Princeton University Press.

Bremer, K. and H.-E. Wanntorp. 1979. Geographic populations or biological species in phylogeny reconstruction? *Systematic Zoology* 28:220–224.

Brooks, D. R. 1985. Historical ecology: a new approach to studying the evolution of ecological associations. *Annals of the Missouri Botanical Garden* 72:660–680.

Brooks, D. R. and D. A. McLennan. 1991. *Phylogeny, Ecology, and Behavior. A Research Program in Comparative Biology*. Chicago: University of Chicago Press.

Brooks, D. R. and E. O. Wiley. 1988. *Evolution as Entropy. Towards a Unified Theory of Biology*. 2d ed. Chicago: University of Chicago Press.

Brothers, D. J. 1985. Species concepts, speciation, and higher taxa. In: E. S. Vrba, ed. *Species and Speciation*, monograph no. 4., pp. 35–42. Pretoria, North Africa: Transvaal Museum.

Brundin, L. 1966. *Transantarctic Relationships and Their Significance, as Evidenced by Chironomid Midges*. Stockholm: Almquist & Wiksell.

Budd, A. F. and B. D. Mishler. 1990. Species and evolution in clonal organisms—summary and discussion. *Systematic Botany* 15:166–171.

Buffon, G. L. L. de. 1749–1804. *Histoire Naturelle, Générale et Particulière, avec la Description du Cabinet du Roi* (44 volumes). Paris: De l'Imprimerie Royale.

Burr, B. M. and R. C. Cashner. 1983. *Campostoma pauciradii*, a new cyprinid fish from southeastern United States, with a review of related forms. *Copeia* 1983:101–116.

Carpenter, J. M. and K. C. Nixon. 1994. On outgroups. *Cladistics* 9:413–426.

Cashner, R. C., J. S. Rogers, and J. M. Grady. 1992. Phylogenetic studies of the genus *Fundulus*. In: R. L. Mayden, ed. *Systematics, Historical Ecology, and North American Freshwater Fishes*, pp. 421–452. Stanford, CA: Stanford University Press.

Chesser, R. T. and R. M. Zink. 1994. Modes of speciation in birds: a test of Lynch's (1989) method. *Evolution* 48:490–497.

Claridge, M. A., H. A. Dawah, and M. R. Wilson, eds. 1997. *Species: The Units of Diversity*. London: Chapman & Hall.

Cole, C. J. 1985. Taxonomy of parthenogenetic species of hybrid origin. *Systematic Zoology* 34:359–363.

Collins, J. T. 1991. Viewpoint: a new taxonomic arrangement for some North American amphibians and reptiles. *Herpetological Review* 22:42–43.

Collins, J. T. 1992. The evolutionary species concept: a reply to Van Devender et al. and Montanucci. *Herpetological Review* 23:43–46.

Collins, J. T. 1993. *Amphibians and Reptiles in Kansas.* 3d ed., revised. Public Education Series no. 13. Lawrence, KS: University of Kansas Museum of Natural History.

Coyne, J. A., H. A. Orr, and D. J. Futuyma. 1988. Do we need a new species concept? *Systematic Zoology* 37:190–200.

Cracraft, J. 1974. Phylogenetic models and classification. *Systematic Zoology* 23: 494–521.

Cracraft, J. 1983. Species concepts and speciation analysis. *Current Ornithology* 1: 159–187.

Cracraft, J. 1987. Species concepts and the ontology of evolution. *Biology and Philosophy* 2:329–346.

Cracraft, J. 1989a. Speciation and its ontology: the empirical consequences of alternative species concepts for understanding patterns and processes of differentiation. In: D. Otte and J. A. Endler, eds. *Speciation and Its Consequences*, pp. 28–59. Sunderland, England: Sinauer Associates.

Cracraft, J. 1989b. Species as entities of biological theory. In: M. Ruse, ed. *What the Philosophy of Biology Is,* pp. 31–52. Dordrecht: Kluwer.

Cracraft, J. 1991. Systematics, species concepts, and conservation biology [Abstract]. Presented at the 5th Annual Meeting of the Society of Conservation Biology, Madison, Wisconsin.

Cracraft, J. 1992. The species of the birds-of-paradise (Paradisaeidae): applying the phylogenetic species concept to a complex pattern of diversification. *Cladistics* 8:1–43.

Cracraft, J. 1997. Species concepts in systematics and conservation biology—an ornithological viewpoint. In: M. A. Claridge, H. A. Dawah, and M. R. Wilson, eds. *Species: The Units of Diversity*, pp. 325–339. London: Chapman & Hall.

Darwin, C. 1859. *On the Origin of Species by Means of Natural Selection, or the Preservation of Favored Races in the Struggle for Life.* London: John Murray.

Darwin, F. 1887. *The Life and Letters of Charles Darwin.* London: John Murray.

Davis, J. I. and K. C. Nixon. 1992. Populations, genetic variation, and the delimitation of phylogenetic species. *Systematic Biology* 41:421–435.

De Marais, B. D., T. E. Dowling, M. E. Douglas, W. L. Minkley, and P. C. Marsh. 1992. Origin of *Gila seminuda* (Teleostei: Cyprinidae) through introgressive hybridization: implications for evolution and conservation. *Proceedings of the National Academy of Sciences, U.S.A.* 89:2747–2751.

De Queiroz, K. 1988. Systematics and the Darwinian revolution. *Philosophy of Science* 55:238–259.

De Queiroz, K. and M. J. Donoghue. 1988. Phylogenetic systematics and the species problem. *Cladistics* 4:317–338.

De Queiroz, K. and M. J. Donoghue. 1990a. Phylogenetic systematics or Nelson's version of cladistics? *Cladistics* 6:61–75.

De Queiroz, K. and M. J. Donoghue. 1990b. Phylogenetic systematics and species revisited. *Cladistics* 6:83–90.

De Queiroz, K. and J. Gauthier. 1992. Phylogenetic taxonomy. *Annual Review of Ecology and Systematics* 23:449–480.

De Queiroz, K. and J. Gauthier. 1994. Towards a phylogenetic system of biological nomenclature. *Trends in Ecology and Evolution* 9:27–31.

Dobzhansky, T. 1935. A critique of the species concept in biology. *Philosophy of Science* 2:344–355.

Dobzhansky, T. 1937. *Genetics and the Origin of Species.* New York: Columbia University Press.

Dobzhansky, T. 1950. Mendelian populations and their evolution. *American Naturalist* 84:401–418.

Dobzhansky, T. 1970. *Genetics of the Evolutionary Process.* New York: Columbia University Press.

Dobzhansky, T. and E. Mayr. 1944. Experiments on sexual isolation in *Drosophila.* I. Geographic strains of *Drosophila willistoni. Proceedings of the National Academy of Sciences, U.S.A.* 30:238–244.

Donoghue, M. J. 1985. A critique of the biological species concept and recommendations for a phylogenetic alternative. *Bryologist* 88:172–181.

Donoghue, M. J. and J. W. Kadereit. 1992. Walter Zimmermann and the growth of phylogenetic theory. *Systematic Biology* 41:74–85.

Dowling, T. E. and B. D. De Marais. 1993. Evolutionary significance of introgressive hybridization in cyprinid fishes. *Nature* 362:444–446.

Doyen, J. T. and C. N. Slobodchikoff. 1974. An operational approach to species classification. *Systematic Zoology* 23:239–247.

Echelle, A. A. and A. F. Echelle. 1992. Mode and pattern of speciation in the evolution of inland pupfishes in the *Cyprinodon variegatus* complex (Teleostei: Cyprinodontidae): an ancestor-descendant hypothesis. In: R. L. Mayden, ed. *Systematics, Historical Ecology, and North American Freshwater Fishes,* pp. 691–709. Stanford, CA: Stanford University Press.

Echelle, A. A., A. F. Echelle, and C. D. Crozier. 1983. Evolution of an all female fish, *Menidia clarkhubbsi* (Atherinidae). *Evolution* 37:772–784.

Echelle, A. A. and D. T. Mosier. 1982. *Menidia clarkhubbsi* new species (Pices Atherinidae) an all female species. *Copeia* 1982:533–540.

Ehrendorfer, F. 1984. Artbegriff und Artbildung in botanischer Sicht. *Zeitschrift für zoologische Systematik und Evolutionsforschung* 22: 234–263.

Eldredge, N. 1989. *Macroevolutionary Dynamics.* New York: McGraw-Hill.

Eldredge, N. 1993. What, if anything, is a species? In: W. H. Kimbel and L. B. Martin, eds. *Species, Species Concepts, and Primate Evolution,* pp. 3–20. New York: Plenum.

Eldredge, N. and J. Cracraft. 1980. *Phylogenetic Patterns and the Evolutionary Process.* New York: Columbia University Press.

Emerson, A. E. 1945. Taxonomic categories and population genetics. *Entomological News* 56:14–19.

Endler, J. A. 1989. Conceptual and other problems in speciation. In: D. Otte and J. A. Endler, eds. *Speciation and Its Consequences,* pp. 625–648. Sunderland, England: Sinauer Associates.

Englemann, G. F. and E. O. Wiley. 1977. The place of ancestor-descendant relationships in phylogenetic reconstruction. *Systematic Zoology* 26:1–11.

Faith, D. P. 1992a. Conservation evaluation and phylogenetic diversity. *Biological Conservation* 61:1–10.

Faith, D. P. 1992b. Systematics and conservation: on predicting the feature diversity of subsets of taxa. *Cladistics* 8:361–373.

Farris, J. S. 1982. Outgroups and parsimony. *Systematic Zoology* 31:328–334.

Farris, J. S. 1983. The logical basis of phylogenetic analysis. In: N. Platnick and V. Funk, eds. *Advances in Cladistics,* vol. 2, pp. 7–36. New York: Columbia University Press.

Frey, J. K. 1993. Models of peripheral isolate formation and speciation. *Systematic Biology* 42:373–381.

Frost, D. R. and D. M. Hillis. 1990. Species concepts and practice: herpetological applications. *Herpetologica* 46:87–104.

Frost, D. R. and A. G. Kluge. 1994. A consideration of epistemology in systematic biology, with special reference to species. *Cladistics* 10:259–294.

Frost, D. R., A. G. Kluge, and D. M. Hillis. 1992. Species in contemporary herpetology: comments on phylogenetic inference and taxonomy. *Herpetological Review* 23:46–54.

Frost, D. R. and J. W. Wright. 1988. The taxonomy of uniparental species, with special reference to parthenogenetic *Cnemidophorus. Systematic Zoology* 37:200–209.

Funk, V. A. 1985. Phylogenetic patterns and hybridization. *Annals of the Missouri Botanical Garden* 72:681–715.

Funk, V. A. and D. R. Brooks. 1990. Phylogenetic systematics as the basis of comparative biology. *Smithsonian Contributions to Botany* 73:1–45.

Futuyma, D. J. 1986. *Evolutionary Biology.* 2d ed. Sunderland, England: Sinauer Associates.

Geist, V. 1992. Endangered species and the law. *Nature* 357:274–276.

Ghiselin, M. T. 1974a. A radical solution to the species problem. *Systematic Zoology* 23:536–544.

Ghiselin, M. T. 1974b. *The Economy of Nature and the Evolution of Sex.* Berkeley, CA: University of California Press.

Ghiselin, M. T. 1981. Categories, life, and thinking. *Behavioral and Brain Sciences* 4:269–313.

Gingerich, P. D. 1979. Populations, phylogeny, and classification: an example from the mammalian research. *Systematic Zoology* 28:451–464.

Grady, J. M. and W. H. LeGrande. 1992, Phylogenetic relationships, modes of speciation, and historical biogeography of the madtom catfishes, genus *Noturus* Rafinesque (Siluriformes: Ictaluridae). In: R. L. Mayden, ed. *Systematics, Historical Ecology, and North American Freshwater Fishes,* pp. 747–777. Stanford, CA: Stanford University Press.

Grant, B. R. and P. R. Grant. 1989. *Evolutionary Dynamics of a Natural Population. The Large Cactus Finch of the Galapagos.* Chicago: University of Chicago Press.

Grant, V. 1994. Evolution of the species concept. *Biologisches Zentralblatt* 113:401–415.

Grant, W. S. 1995. Multi-disciplinary approaches for defining evolutionarily significant units for conservation. *South African Journal of Science* 91:65–67.

Greenwood, P. H. 1984. African cichlids and evolutionary theories. In: A. A. Echelle and I. Cornfield, eds. *Evolution of Fish Species Flocks,* pp. 141–154. Orono, ME: Orono Press.

Griffiths, G. C. D. 1974. On the foundations of biological systematics. *Acta Biotheoretica* 23:85–131.

Hacking, I. 1991. A tradition of natural kinds. *Philosophical Studies* 61:109–126.

Harper, C. W. Jr. 1976. Phylogenetic inference in paleontology. *Journal of Paleontology* 50:180–193.

Harper, J. L. 1977. *Population Biology of Plants.* London: Academic Press.

Haszprunar, G. 1992. The types of homology and their significance for evolutionary biology and phylogenetics. *Journal of Evolutionary Biology* 5:13–24.

Häuser, C. L. 1987. The debate about the biological species concept—a review. *Zeitschrift für zoologische Systematik und Evolutionsforschung* 25:241–257.

Hengeveld, R. 1988. Mayr's ecological species criterion. *Systematic Zoology* 37:47–55.

Hennig, W. 1950. *Grundzüge einer Theorie der phylogenetischen Systematik.* Berlin: Aufbau.

Hennig, W. 1953. Kritische Bemerkungen zum phylogenetischen System der Insekten. *Beiträge zur Entomologie* 3:1–61.

Hennig, W. 1957. Systematik und Phylogenese. *Bericht Hundertjahrfeier Deutsche Entomologische Gesellschaft* 1956:51–71.

Hennig, W. 1965. Phylogenetic systematics. *Annual Review of Entomology* 10:97–116.

Hennig, W. 1966. *Phylogenetic Systematics.* Urbana, IL: University of Illinois Press.

Hennig, W. 1969. *Die Stammesgeschichte der Insekten.* Frankfurt am Main: Waldemar Kramer.

Hennig, W. 1981. *Insect Phylogeny.* Chichester and New York: Wiley.

Hennig, W. 1982. *Phylogenetische Systematik.* Berlin: Paul Parey.

Hennig, W. 1984. *Aufgaben und Probleme stammesgeschichtlicher Forschung.* Berlin: Paul Parey.

Hennig, W. and D. Schlee. 1978. Abriß der phylogenetischen Systematik. *Stuttgarter Beiträge zur Naturkunde, Serie A* 319:1–11.

Hill, C. R. and P. R. Crane. 1982. Evolutionary cladistics and the origin of angiosperms. In: K. A. Joysey and A. E. Friday, eds. *Problems of Phylogenetic Reconstruction,* pp. 269–361. New York: Academic Press.

Holman, E. W. 1987. Recognizability of sexual and asexual species of rotifers. *Systematic Zoology* 36:381–386.

Horvath, C. D. 1997. Some questions about identifying individuals: failed intuitions about organisms and species. *Philosophy of Science* 64:654–668.

Hubbs, C. 1967. Analysis of phylogenetic relationships using hybridization techniques. *Bulletin of the National Institute of Sciences, India* 34:48–59.

Hubbs, C. L. 1955. Hybridization between fish species in nature. *Systematic Zoology* 4:1–20.

Hull, D. L. 1964. Consistency and monophyly. *Systematic Zoology* 13:1–11.

Hull, D. L. 1968. The operational imperative—sense and nonsense in operationalism. *Systematic Zoology* 17:438–457.

Hull, D. L. 1979. The limits of cladism. *Systematic Zoology* 28:416–440.

Hull, D. L. 1980. Individuality and selection. *Annual Review of Ecology and Systematics* 11:311–332.

Hull, D. L. 1988. *Science as a Process. An Evolutionary Account of the Social and Conceptual Development of Science*. Chicago: University of Chicago Press.

Hull, D. L. 1997. The ideal species concept--and why we can't get it. In: M. A. Claridge, H. A. Dawah, and M. R. Wilson, eds. *Species: The Units of Diversity*, pp. 357–380. London: Chapman & Hall.

Hunt, D. J. 1997. Nematode species: concepts and identification strategies exemplified by the Longidoridae, Steinernematidae, and Heterorhabditidae. In: M. A. Claridge, H. A. Dawah, and M. R. Wilson, eds. *Species: The Units of Diversity*, pp. 221–245. London: Chapman & Hall.

Huxley, J. S. 1940. Towards the new systematics. In: J. S. Huxley, ed. *The New Systematics*, pp. 1–46. London: Clarendon.

Johnson, L. A. S. 1970. Rainbow's end: the quest for an optimal taxonomy. *Systematic Zoology* 19:203–239.

Key, K. H. 1981. Species, parapatry, and the morbine grasshoppers. *Systematic Zoology* 30:425–458.

Kimbel, W. H. and L. B. Martin. 1993. Species and speciation. Conceptual issues and their relevance for primate evolutionary biology. In: W. H. Kimbel and L. B. Martin, eds. *Species, Species Concepts, and Primate Evolution*, pp. 539–553. New York: Plenum.

Kimbel, W. H. and Y. Rak. 1993. The importance of species taxa in paleoanthropology and an argument for the phylogenetic concept of the species category. In: W. H. Kimbel and L. B. Martin, eds. *Species, Species Concepts, and Primate Evolution*, pp. 461–484. New York: Plenum.

Klausnitzer, B. and K. Richter. 1979. Bemerkungen zum Artkonzept und zur Phylogenie der Arten. *Zeitschrift für zoologische Systematik und Evolutionsforschung* 17:236–241.

Kluge, A. G. 1989. Metacladistics. *Cladistics* 5:291–294.

Kluge, A. G. 1990. Species as historical individuals. *Biology and Philosophy* 5:417–431.

Königsmann, E. 1975. Termini der phylogenetischen Systematik. *Biologische Rundschau* 13:99–115.

Kottler, M. 1978. Charles Darwin's biological species concept and theory of geographic speciation: the transmutation notebooks. *Annals of Science* 35:275–297.

Krishtalka, L. 1993. Anagenetic angst. Species boundaries in Eocene primates. In: W. H. Kimbel and L. B. Martin, eds. *Species, Species Concepts, and Primate Evolution*, pp. 331–344. New York: Plenum.

Lane, R. 1997. The species concept in blood-sucking vectors of human disease. In: M. A. Claridge, H. A. Dawah, and M. R. Wilson, eds. *Species: The Units of Diversity*, pp. 273–289. London: Chapman & Hall.

Lauterbach, K.-E. 1992. Phylogenetische Systematik. *Berichte des Naturwissenschaftlichen Vereins Bielefeld* 33:209–240.

Lewontin, R. C. 1974. *The Genetic Basis of Evolutionary Change*. New York: Columbia University Press.

Liden, M. 1990. Replicators, hierarchy, and the species problem. *Cladistics* 6:183–186.

Lynch, J. D. 1989. The gauge of speciation: on the frequencies of modes of speciation. In: D. Otte and J. A. Endler, eds. *Speciation and Its Consequences*, pp. 527–553. Sunderland, England: Sinauer Associates.

Markle, D. F., T. N. Pearsons, and D. T. Bills. 1991. Natural history of *Oregonichthys* (Pisces: Cyprinidae), with a description of a new species from the Umpqua River of Oregon. *Copeia* 1991:277–293.

Mayden, R. L. 1985. Biogeography of Ouachita Highlands fishes. *Southwestern Naturalist* 30:195–211.

Mayden, R. L. 1987. Historical ecology and North American highlands fishes: a research program in community ecology. In: W. J. Matthews and D. C. Heins, eds. *Community and Evolutionary Ecology of North American Stream Fishes,* pp. 210–222. Norman, OK: University of Oklahoma Press.

Mayden, R. L. 1988. Vicariance biogeography, parsimony, and evolution in North American freshwater fishes. *Systematic Zoology* 37:331–357.

Mayden, R. L. 1997. A hierarchy of species concepts: the denouement in the saga of the species problem. In: M. A. Claridge, H. A. Dawah, and M. R. Wilson, eds. *Species: The Units of Diversity,* pp. 381–424. London: Chapman & Hall.

Mayden, R. L., B. M. Burr, L. M. Page, and R. R. Miller. 1992. The native fishes of North America. In: R. L. Mayden, ed. *Systematics, Historical Ecology, and North American Freshwater Fishes,* pp. 827–863. Stanford, CA: Stanford University Press.

Mayden, R. L. and B. R. Kuhajda. 1989. Systematics of *Notropis cahabae,* a new cyprinid fish endemic to the Cahaba River of the Mobile Basin. *Bulletin of the Alabama Museum of Natural History,* No. 9.

Mayden, R. L. and R. H. Matson. 1992. Systematics and biogeography of the Tennessee shiner, *Notropis leuciodus* (Cope) (Teleostei: Cyprinidae). *Copeia* 1992:954–968.

Mayden, R. L. and E. O. Wiley. 1992. The fundamentals of phylogenetic systematics. In: R. L. Mayden, ed. *Systematics, Historical Ecology, and North American Freshwater Fishes,* pp. 114–185. Stanford, CA: Stanford University Press.

Maynard-Smith, J. 1986. *The Problems of Biology.* Oxford: Oxford University Press.

Mayr, E. 1942. *Systematics and the Origin of Species from the Viewpoint of a Zoologist.* New York: Columbia University Press.

Mayr, E. 1946a. Experiments on sexual isolation in *Drosophila.* VI. Isolation between *Drosophila pseudoobscura* and *Drosophila persimilis* and their hybrids. *Proceedings of the National Academy of Sciences, U.S.A.* 32:57–59.

Mayr, E. 1946b. Experiments on sexual isolation in *Drosophila.* VII. The nature of the isolating mechanisms between *Drosophila pseudoobscura* and *Drosophila persimilis. Proceedings of the National Academy of Sciences, U.S.A.* 32:128–137.

Mayr, E. 1948. The bearing of the New Systematics on genetical problems; the nature of species. *Advances in Genetics* 2:205–237.

Mayr, E. 1950. The role of the antennae in the mating behavior of female *Drosophila. Evolution* 4:149–154.

Mayr, E. 1957a. Species concepts and definitions. In: E. Mayr, ed. *The Species Problem,* publication no. 50, pp. 1–22. Washington, DC: American Association for the Advancement of Science.

Mayr, E. 1957b. Difficulties and importance of the biological species concept. In: E. Mayr, ed. *The Species Problem,* publication no. 50, pp. 371–388. Washington, DC: American Association for the Advancement of Science.

Mayr, E. 1963. *Animal Species and Evolution.* Cambridge: Belknap Press.

Mayr, E. 1969. *Principles of Systematic Zoology.* New York: McGraw-Hill.

Mayr, E. 1970. *Populations, Species, and Evolution.* Cambridge: Harvard University Press.

Mayr, E. 1974. Cladistic analysis or cladistic classification. *Zeitschrift für zoologische Systematik und Evolutionsforschung* 12:94–128.

Mayr, E. 1976. Agassiz, Darwin, and evolution. In: E. Mayr, ed. *Evolution and the Diversity of Life. Selected Essays,* pp. 251–276. Cambridge: Belknap Press.

Mayr, E. 1982. *The Growth of Biological Thought: Diversity, Evolution, and Inheritance.* Cambridge: Belknap Press.

Mayr, E. 1988a. *Toward a New Philosophy of Biology: Observations of an Evolutionist.* Cambridge, MA: Harvard University Press.

Mayr, E. 1988b. The why and how of species. *Biology and Philosophy* 3:431–441.

Mayr, E. 1988c. Recent historical developments. In: D. L. Hawksworth, ed. *Prospects in Systematics,* pp. 31–43. Oxford: Clarendon.

Mayr, E. 1992a. A local flora and the biological species concept. *American Journal of Botany* 79:222–238.

Mayr, E. 1992b. The principle of divergence. *Journal of the History of Biology* 25: 343–359.

Mayr, E. 1995. Systems of ordering data. *Biology and Philosophy* 10:419–434.

Mayr, E. 1996. What is a species and what is not? *Philosophy of Science* 63:262–277.

Mayr, E. 1997. Perspective: the objects of selection. *Proceedings of the National Academy of Sciences, U.S.A.* 94:2091–2094.

Mayr, E. and P. D. Ashlock. 1991. *Principles of Systematic Zoology.* 2d ed. New York: McGraw-Hill.

Mayr, E. and T. Dobzhansky. 1945. Experiments on sexual isolation in *Drosophila.* IV. Modification of the degree of isolation between *Drosophila pseudoobscura* and *Drosophila persimilis* and of sexual preferences in *Drosophila prosaltans. Proceedings of the National Academy of Sciences, U.S.A.* 31:75–82.

Mayr, E. and L. L. Short. 1970. *Species Taxa of North American Birds,* publication no. 9. Cambridge, MA: Nuttall Ornithological Club.

McAllister, D. E. and B. W. Coad. 1978. A test between relationships based on phenetic and cladistic taxonomic methods. *Canadian Journal of Zoology* 56:2198–2210.

McDade, L. A. 1992. Hybrids and phylogenetic systematics II. The impact of hybrids on cladistic analysis. *Evolution* 46:1329–1346.

McKitrick, M. 1994. On homology and the ontological relationship of parts. *Systematic Biology* 43:1–10.

Meffe, G. K. and F. F. Snelson Jr. 1989. *Ecology and Evolution of Livebearing Fishes (Poeciliidae).* Englewood Cliffs, NJ: Prentice-Hall.

Meyer, A. 1990. Ecological and evolutionary aspects of the trophic polymorphism in *Cichlasoma citrinellum* (Pisces: Cichlidae). *Biological Journal of the Linnean Society, London* 39:279–299.

Mishler, B. D. 1985. The morphological, developmental, and phylogenetic basis of species concepts in bryophytes. *Bryologist* 88:207–214.

Mishler, B. D. 1987. Sociology of science and the future of Hennigian phylogenetic systematics. *Cladistics* 3:55–60.

Mishler, B. D. 1990. Reproductive biology and species distinctions in the moss genus *Tortula*, as represented in Mexico. *Systematic Botany* 15:86–97.

Mishler, B. D. 1994. The cladistic analysis of molecular and morphological data. *American Journal of Physical Anthropology* 94:143–156.

Mishler, B. D. 1999. Getting rid of species? In: R. Wilson, ed. *Species: New Interdisciplinary Essays,* pp. 307–315. Cambridge: MIT Press.

Mishler, B. D. and R. N. Brandon.. 1987. Individuality, pluralism, and the phylogenetic species concept. *Biology and Philosophy* 2:397–414.

Mishler, B. D. and A. F. Budd. 1990. Species and evolution in clonal organisms—introduction. *Systematic Botany* 15:79–85.

Mishler, B. D. and M. J. Donoghue. 1982. Species concepts: a case for pluralism. *Systematic Zoology* 31:491–503.

Montanucci, R. R. 1992. Commentary on a proposed taxonomic arrangement of some North American amphibians and reptiles. *Herpetological Review* 23:9–10.

Moritz, C. 1994a. Applications of mitochondrial DNA analysis in conservation: a critical review. *Molecular Ecology* 3:401–411.

Moritz, C. 1994b. Defining "evolutionarily significant units" for conservation. *Trends in Ecology and Evolution* 9:373–375.

Moritz, C. 1995. Uses of molecular phylogenies for conservation. *Philosophical Transactions of the Royal Society of London* 349B:113–118.

Naef, A. 1919. *Idealistische Morphologie und Phylogenetik.* Jena, Germany: Gustav Fischer.

Nelson, G. J. 1971. Paraphyly and polyphyly: redefinitions. *Systematic Zoology* 20:471–472.

Nelson, G. 1973. Classification as an expression of phylogenetic relationships. *Systematic Zoology* 22:344–359.

Nelson, G. J. 1989a. Species and taxa: systematics and evolution. In: D. Otte and J. A. Endler, eds. *Speciation and Its Consequences,* pp. 60–81. Sunderland, England: Sinauer Associates.

Nelson, G. J. 1989b. Cladistics and evolutionary models. *Cladistics* 5:275–289.

Nelson, G. and C. Patterson. 1993. Cladistics, sociology and success: a comment on Donoghue's critique of David Hull. *Biology and Philosophy* 8:441–443.

Nelson, G. J. and N. I. Platnick. 1981. *Systematics and Biogeography: Cladistics and Vicariance.* New York: Columbia University Press.

Nixon, K. C. and Q. D. Wheeler. 1990. An amplification of the phylogenetic species concept. *Cladistics* 6:211–223.

Nixon, K. C. and Q. D. Wheeler. 1992a. Extinction and the origin of species. In: M. J. Novacek and Q. D. Wheeler, eds. *Extinction and Phylogeny,* pp. 119–143. New York: Columbia University Press.

Nixon, K. C. and Q. D. Wheeler. 1992b. Measures of phylogenetic diversity. In: M. J. Novacek and Q. D. Wheeler, eds. *Extinction and Phylogeny,* pp. 216–234. New York: Columbia University Press.

Norris, S. M. and M. E. Douglas. 1992. Geographic variation, taxonomic status, and biogeography of two widely distributed African freshwater fishes, *Ctenopoma petherici* and *C. kingsleyae* (Teleostei: Anabantidae). *Copeia* 1992:702–724.

Novacek, M. J. and Q. D. Wheeler. 1992. Introduction. Extinct taxa: accounting for 99.999 . . . % of the Earth's biota. In: M. J. Novacek and Q. D. Wheeler, eds. *Extinction and Phylogeny,* pp. 1–16. New York: Columbia University Press.

O'Hara, R. J. 1993. Systematic generalization, historical fate, and the species problem. *Systematic Biology* 42:231–246.

Osche, G. 1984. Artbegriff und Artbildung in zoologischer, botanischer, und paläontologischer Sicht—Einleitung. *Zeitschrift für zoologische Systematik und Evolutionsforschung* 22:164–168.

Page, L. M. and B. M. Burr. 1991. *A Field Guide to Freshwater Fishes. North America North of Mexico.* Boston: Houghton Mifflin.

Paterson, H. E. H. 1973. Animal species studies. *Journal of the Royal Society of Western Australia* 36:31–36.

Paterson, H. E. H. 1978. More evidence against speciation by reinforcement. *South African Journal of Science* 74:369–371.

Paterson, H. E. H. 1980. A comment on mate recognition systems. *Evolution* 34:330–331.

Paterson, H. E. H. 1981. The continuing search for the unknown and the unknowable. *South African Journal of Science* 77:113–119.

Paterson, H. E. H. 1982. Perspective on speciation by reinforcement. *South African Journal of Science* 78:53–57.

Paterson, H. E. H. 1985. The recognition concept of species. In: E. S. Vrba, ed. *Species and Speciation,* monograph no. 4, pp. 21–29. Pretoria: Transvaal Museum.

Paterson, H. E. H. 1986. Environment and species. *South African Journal of Science* 82:62–65.

Paterson, H. E. H. 1993. *Evolution and the Recognition Concept of Species.* Baltimore: Johns Hopkins University Press.

Patterson, C. 1982. Morphological characters and homology. In: K. A. Joysey and A. E. Friday, eds. *Problems of Phylogenetic Reconstruction,* pp. 21–74. London: Academic.

Patterson, C. and A. B. Smith. 1987. Is the periodicity of extinctions a taxonomic artefact? *Nature* 330:248–251.

Patton, J. L. and M. F. Smith. 1994. Paraphyly, polyphyly, and the nature of species boundaries in pocket gophers (Genus *Thomomys*). *Systematic Biology* 43:11–26.

Peters, D. S. 1970. Über den Zusammenhang von biologischem Artbegriff und phylogenetischer Systematik. *Aufsätze und Reden der Senckenbergischen naturforschenden Gesellschaft* 18:1–39.

Plate, L. 1914. Prinzipien der Systematik mit besonderer Berücksichtigung des Systems der Tiere. *Kultur der Gegenwart* 3:119–159.

Platnick, N. I. 1977a. [Review of *Concepts of Species,* C. N. Slobodchikoff, ed. New York: Wiley, 1976.] *Systematic Zoology* 26:96–98.

Platnick, N. I. 1977b. Monophyly and the origin of higher taxa: a reply to E. O. Wiley. *Systematic Zoology* 26:355–357.

Platnick, N. I. 1977c. Cladograms, phylogenetic trees, and hypothesis testing. *Systematic Zoology* 26:438–442.

Platnick, N. I. 1979. Philosophy and the transformation of cladistics. *Systematic Zoology* 28:537–546.

Prothero, D. R. and D. B. Lazarus. 1980. Planktonic microfossils and the recognition of ancestors. *Systematic Zoology* 28:119–129.

Raubenheimer, D. and T. M. Crowe. 1987. The recognition concept: is it really an alternative? *South African Journal of Science* 83:530–534.

Raup, D. M. and J. Sepkoski, Jr. 1986. Periodic extinctions of families and genera. *Science* 231:833–836.

Raven, P. H. et al. 1993. *A Biological Survey for the Nation*. Washington, DC: National Academy Press.

Regan, C. T. 1926. Organic evolution. *Report of the British Association for the Advancement of Science* 1925:75–86.

Rensch, B. 1959. *Evolution above the Species Level.* New York: Columbia University Press.

Richter, S. and R. Meier. 1994. The development of phylogenetic concepts in Hennig's early theoretical publications (1947–1966). *Systematic Biology* 43:212–221.

Ridley, M. 1986. *Evolution and Classification: The Reformation of Cladism.* London: Longman.

Ridley, M. 1989. The cladistic solution to the species problem. *Biology and Philosophy* 4:1–16.

Ridley, M. 1993. *Evolution*. Boston: Blackwell Scientific Publications.

Riedl, R. 1978. *Order in Living Organisms.* New York: Wiley-Interscience.

Rieppel, O. 1988. *Fundamentals of Comparative Biology.* Basel: Birkhauser.

Rieseberg, L. H. 1991. Homoploid reticulate evolution in *Helianthus* (Asteraceae): evidence from ribosomal genes. *American Journal of Botany* 78:1218–1237.

Rieseberg, L. H., R. Carter, and S. Zona. 1990. Molecular tests of the hypothesized hybrid origin of two diploid *Helianthus* species (Asteraceae). *Evolution* 44:1498–1511.

Robins, C. R. (chairman of the Committee on Names of Fishes). 1980. *A List of Common and Scientific Names of Fishes from the United States and Canada.* 4th ed. Special publication no. 12. Washington, DC: American Fisheries Society.

Robins, C. R. (chairman of the Committee on Names of Fishes). 1991. *Common and Scientific Names of Fishes from the United States and Canada.* 5th ed. Special publication no. 20. Washington, DC: American Fisheries Society.

Rose, K. D. and T. M. Brown. 1993. Species concepts and species recognition in Eocene primates. In: W. H. Kimbel and L. B. Martin, eds. *Species, Species Concepts, and Primate Evolution,* pp. 299–330. New York: Plenum.

Rosen, D. E. 1978. Vicariant patterns and historical explanation in biogeography. *Systematic Zoology* 27:159–188.

Rosen, D. E. 1979. Fishes from the uplands and intermontane basics of Guatemala: revisionary studies and comparative biogeography. *Bulletin of the American Museum of Natural History* 162:267–376.

Rosenberg, A. 1985. *The Structure of Biological Science.* New York: Cambridge University Press.

Ross, H. H. 1972. The origin of species diversity in ecological communities. *Taxon* 21:253–259.

Ryder, O. A. 1986. Species conservation and systematics: the dilemma of subspecies. *Trends in Ecology and Evolution* 1:9–10.

Signor, P. W. III. 1985. Real and apparent trends in species richness through time. In: J. W. Valentine, ed. *Phanerozoic Diversity Patterns: Profiles in Macroevolution*, pp. 129–150. Princeton, NJ: Princeton University Press.

Simpson, G. G. 1944. The principles of classification and a classification of mammals. *Bulletin of the American Museum of Natural History* 85:1–350.

Simpson, G. G. 1951. The species concept. *Evolution* 5:285–298.

Simpson, G. G. 1953. *The Major Features of Evolution*. New York: Columbia University Press.

Simpson, G. G. 1961. *Principles of Animal Taxonomy*. New York: Columbia University Press.

Sloan, P. 1987. Buffon's species concept. From logical universals to historical individuals: Buffon's idea of biological species. In: J. Roger and F. L. Fischer, eds. *Histoire du Concept d'Espèce dans les Sciences de la Vie*, pp. 101–140. Paris: Fondation Singer-Polignac.

Smith, G. R. 1992. Introgression in fishes: significance for paleontology, cladistics, and evolutionary rates. *Systematic Biology* 41:41–57.

Sneath, P. H. A. and R. R. Sokal. 1973. *Numerical Taxonomy*. San Francisco: Freeman.

Sober, E. 1988. *Reconstructing the Past*. Cambridge, MA: MIT Press.

Sokal, R. R. and T. J. Crovello. 1970. The biological species concept: a critical evaluation. *American Naturalist* 104:127–153.

Sonneborn, T. M. 1975. The *Paramecium aurelia* complex of fourteen sibling species. *Transactions of the American Microscopical Society* 94:155–178.

Sudhaus, W. 1984. Artbegriff und Artbildung in zoologischer Sicht. *Zeitschrift für zoologische Systematik und Evolutionsforschung* 22:183–211.

Sudhaus, W. and K. Rehfeld. 1992. *Einführung in die Phylogenetik und Systematik*. Stuttgart: Gustav Fischer.

Systematics Agenda 2000. 1994a. *Systematics Agenda 2000: Charting the Biosphere*. New York: Society of Systematic Biologists, American Society of Plant Taxonomists, Willi Hennig Society, Association of Systematics Collections.

Systematics Agenda 2000. 1994b. *Systematics Agenda 2000: Charting the Biosphere*, Technical Report. New York: Society of Systematic Biologists, American Society of Plant Taxonomists, Willi Hennig Society, Association of Systematics Collections.

Szalay, F. S. 1993. Species concepts. The tested, the untestable, and the redundant. In: W. H. Kimbel and L. B. Martin, eds. *Species, Species Concepts, and Primate Evolution*, pp. 21–41. New York: Plenum.

Templeton, A. R. 1989. The meaning of species and speciation: a genetic perspective. In: D. Otte and J. A. Endler, eds. *Speciation and Its Consequences*, pp. 3–27. Sunderland, England: Sinauer Associates.

Templeton, A. R. 1994. The role of molecular genetics in speciation studies. In: B. Schierwater et al., eds. *Molecular Ecology and Evolution*, pp. 455–477. Basel, Switzerland: Birkhauser.

Theriot, E. 1992. Clusters, species concepts, and morphological evolution of diatoms. *Systematic Biology* 41:141–157.

Van Devender, T. R., C. H. Lowe, H. K. McCrystal, and H. E. Lawler. 1992. Viewpoint:

reconsider suggested systematic arrangements for some North American amphibians and reptiles. *Herpetological Review* 23:10–14.

Van Valen, L. 1976. Ecological species, multispecies, and oaks. *Taxon* 25:233–239.

Vane-Wright, R. I., C. J. Humphries, and P. H. Williams. 1991. What to protect?—Systematics and the agony of choice. *Biological Conservation* 55:235–254.

Venn, J. 1866. *The Logic of Chance. An Essay on the Foundations and Province of the Theory of Probability, with Especial Reference to Its Application to Moral and Social Science.* London: Macmillan.

Vogler, A. P. and DeSalle, R. 1994. Diagnosing units of conservation management. *Conservation Biology* 8:354–363

Vrana, P. and W. Wheeler. 1992. Individual organisms as terminal entities: laying the species problem to rest. *Cladistics* 8:67–72.

Wagner, G. P. 1989. The biological homology concept. *Annual Review of Ecology and Systematics* 20:51–69.

Walker, J. M. 1986. The taxonomy of parthenogenetic species of hybrid origin: cloned hybrid populations of *Cnemidophorus* (Sauria: Teiidae). *Systematic Zoology* 35:427–440.

Wallace, A. R. 1855. On the law which has regulated the introduction of new species. *Annals and Magazine of Natural History* 16:184–196.

Wheeler, Q. D. 1993. A crisis of biodiversity: systematics and ecology [review of *Systematics, Ecology, and the Biodiversity Crisis,* N. Eldredge, ed. New York: Columbia University Press, 1992]. *BioScience* 43:578–580.

Wheeler, Q. D. 1995a. Systematics, the scientific basis for inventories of biodiversity. *Biodiversity and Conservation* 4:476–489.

Wheeler, Q. D. 1995b. Systematics and biodiversity: policies at higher levels. *Bioscience, Science and Biodiversity Policy Supplement,* pp. 21–28.

Wheeler, Q. D. 1995c. The "Old Systematics": classification and phylogeny. In: J. Pakaluk and S. A. Slipinski, eds. *Biology and Classification of Coleoptera*, pp. 31–62. Warsaw: Muzeum i Instytut Zoologii PAN.

Wheeler, Q. D. and J. Cracraft. 1996. Taxonomic preparedness: are we ready to meet the biodiversity challenge? In: D. Wilson, M. Reaka-Kudla, and E. O. Wilson, eds. *Biodiversity II.* Washington, DC: National Academy Press.

Wheeler, Q. D. and K. C. Nixon. 1990. Another way of looking at the species problem: a reply to De Queiroz and Donoghue. *Cladistics* 6:77–81.

White, M. J. D. 1978. *Modes of Speciation.* San Francisco: Freeman.

Wiley, E. O. 1975. Karl R. Popper, systematics and classification: a reply to Walter Bock and other evolutionary taxonomists. *Systematic Zoology* 24:233–243.

Wiley, E. O. 1977. The phylogeny and systematics of the *Fundulus nottii* species group (Teleostei: Cyprinodontidae). *Occasional Papers of the Museum of Natural History, University of Kansas* 66:1–31.

Wiley, E. O. 1978. The evolutionary species concept reconsidered. *Systematic Zoology* 27:17–26.

Wiley, E. O. 1979a. Cladograms and phylogenetic trees. *Systematic Zoology* 28:88–92.

Wiley, E. O. 1979b. An annotated Linnean hierarchy, with comments on natural taxa and competing systems. *Systematic Zoology* 28:308–337.

Wiley, E. O. 1979c. Ancestors, species and cladograms—remarks on the symposium. In: J. Cracraft and N. Eldredge, eds. *Phylogenetic Analysis and Paleontology*, pp. 211–225. New York: Columbia University Press.

Wiley, E. O. 1980a. Is the evolutionary species fiction?—A consideration of classes, individuals and historical entities. *Systematic Zoology* 29:76–80.

Wiley, E. O. 1980b. The metaphysics of individuality and its consequences for systematic biology. *Behavioral and Brain Science* 4:302–303.

Wiley, E. O. 1981a. *Phylogenetics. The Theory and Practice of Phylogenetic Systematics.* New York: Wiley-Interscience.

Wiley, E. O. 1981b. Convex groups and consistent classifications. *Systematic Botany* 6:346–358.

Wiley, E. O. 1986. A study of the evolutionary relationships of *Fundulus* topminnows (Teleostei: Fundulidae). *American Zoologist* 26:121–130.

Wiley, E. O. 1987a. The evolutionary basis for phylogenetic classification. In: P. Hovenkamp, ed. *Systematics and Evolution: A Matter of Diversity,* pp. 55–64. Utrecht: Utrecht University.

Wiley, E. O. 1987b. Process and pattern. Cladograms and trees. In: P. Hovenkamp, ed. *Systematics and Evolution: A Matter of Diversity,* pp. 233–247. Utrecht: Utrecht University.

Wiley, E. O. 1989. Kinds, individuals and theories. In: M. Ruse, ed. *What the Philosophy of Biology Is: Essays Dedicated to David Hull,* pp. 289–300. Dordrecht: Kluwer.

Wiley, E. O. and D. D. Hall. 1975. *Fundulus blariae*, a new species of the *Fundulus nottii* complex. *American Museum Novitates* No. 2577:1–14.

Wiley, E. O. and R. L. Mayden. 1985. Species and speciation in phylogenetic systematics, with examples from the North American fish fauna. *Annals of the Missouri Botanical Garden* 72:596–635.

Wiley, E. O., D. Siegel-Causey, D. R. Brooks, and V. A. Funk. 1991. *The Compleat Cladist. A Primer of Phylogenetic Procedures,* special publication no. 19. Lawrence, KS: University of Kansas Museum of Natural History.

Willmann, R. 1981. Evolution, Systematik und stratigraphische Bedeutung der neogenen Süßwassergastropoden von Rhodos und Kos/Ägäis. *Palaeontographica, Abteilung A* 174:10–235.

Willmann, R. 1985a. *Die Art in Raum und Zeit.* Berlin: Paul Parey.

Willmann, R. 1985b. Responses of the Plio-Pleistocene freshwater gastropods of Kos (Greece, Aegean Sea) to environmental changes. In: U. Bayer and A. Seilacher, eds. *Sedimentary and Evolutionary Cycles.* [*Lecture Notes in Earth Sciences* 1:295–321.] New York: Springer.

Willmann, R. 1986. Reproductive isolation and the limits of the species in time. *Cladistics* 2:356–358.

Willmann, R. 1989. Evolutionary or biological species. *Abhandlungen des naturwissenschaftlichen Vereins Hamburg (NF)* 28:95–110.

Willmann, R. 1991. Die Art als Taxon und als Einheit der Natur. *Mitteilungen des Zoologischen Museums Berlin* 67:5–15.

Willmann, R. 1997. Phylogeny and the consequences of phylogenetic systematics. In:

P. Landolt and M. Sartori, eds. *Ephemeroptera and Plecoptera. Biology-Ecology-Systematics,* pp. 499–510. Fribourg, Switzerland: Mauron & Tinguely & Lachat.

Wilson, E. O. 1985. The biological diversity crisis: a challenge to science. *Issues in Science and Technology* fall:20–29.

Wilson, E. O., ed. 1988. *Biodiversity.* Washington, DC: National Academy of Science Press.

Wilson, E. O. 1992. *The Diversity of Life.* Cambridge, MA: Harvard University Press.

Wilson, B. E. 1995. A (not so radical) solution to the species problem. *Biology and Philosophy* 10:339–356.

Wood, R. M. and R. L. Mayden. 1992. Systematics and biogeography of *Notropis lutipinnis* and *N. chlorocephalus. Copeia* 1992:68–81.

Woodger, J. H. 1937. *The Axiomatic Method in Biology.* London: Cambridge University Press.

Woodger, J. H. 1952a. Science without properties. *British Journal for the Philosophy of Science* 2:193–216.

Woodger, J. H. 1952b. From biology to mathematics. *British Journal for the Philosophy of Science* 3:1–21.

Woodger, J. H. 1953. What do we mean by "inborn"? *British Journal for the Philosophy of Science* 3:319–326.

Woodger, J. H. 1961. Taxonomy and evolution. *Nuova Critica* 3:67–78.

Zechman, F. W., E. A. Zimmer, and E. C. Theriot. 1994. Use of ribosomal DNA internal transcribed spacers for phylogenetic studies in diatoms. *Journal of Phycology* 30:507–512.

Zimmermann, W. 1931. Arbeitsweise der botanischen Phylogenetik und anderer Gruppierungswissenschaften. In: E. Abderhalden, ed. *Handbuch der biologischen Arbeitsmethoden,* Abteilung 3.2, Teil 9, pp. 941–1053. Berlin: Urban and Schwarzenberg.

Zimmermann, W. 1943. Die Methoden der Phylogenetik. In: G. Heberer, ed. *Die Evolution der Organismen 1,* Aufl. G., pp. 20–56. Jena, Germany: Gustav Fischer.

INDEX

Numbers in bold indicate pages where figures appear. Page ranges by each author are indicated in header.

Cracraft: 3–14, **Mayr:** 17–29, 93–100, 161–166; **Meier and Willmann:** 30–43, 101–118, 167–178; **Mishler and Theriot:** 44–54, 119–132, 179–184; **Wheeler and Platnick:** 55–69, 133–145, 185–197; **Wiley and Mayden:** 70–89, 146–158, 198–208

Cracraft: 3–14, **Mayr:** 17–29, 93–100, 161–166; **Meier and Willmann:** 30–43, 101–118, 167–178; **Mishler and Theriot:** 44–54, 119–132, 179–184; **Wheeler and Platnick:** 55–69, 133–145, 185–197; **Wiley and Mayden:** 70–89, 146–158, 198–208